Lecture Notes in Control and Information Sciences

Edited by M. Thoma and A. Wyner

Vol. 81: Stochastic Optimization
Proceedings of the International Conference, Kiew, 1984
Edited by I. Arkin, A. Shiraev, R. Wets
X, 754 pages, 1986.

Vol. 82: Analysis and Algorithms of Optimization Problems
Edited by K. Malanowski, K. Mizukami
VIII, 240 pages, 1986.

Vol. 83: Analysis and Optimization of Systems
Proceedings of the Seventh International Conference of Analysis and Optimization of Systems
Antiba, June 26-27, 1986
Edited by A. Bensoussan, J. L. Lions
XVI, 901 pages, 1986.

Vol. 84: System Modelling and Optimization
Proceedings of the 12th IFIP Conference
Budapest, Hungary, September 2–6, 1985
Edited by A. Prékopa, J. Szelezsán, B. Strazicky
XII, 1046 pages, 1986.

Vol. 85: Stochastic Processes in Underwater Acoustics
Edited by Charles R. Baker
V, 205 pages, 1986.

Vol. 86: Time Series and Linear Systems
Edited by Sergio Bittanti
XVII, 243 pages, 1986.

Vol. 87: Recent Advances in System Modelling and Optimization
Proceedings of the IFIP-WG 7/1 Working Conference
Santiago, Chile, August 27-31, 1984
Edited by L. Contesse, R. Correa, A. Weintraub
IV, 199 pages, 1987.

Vol. 88: Bruce A. Francis
A Course in H_∞ Control Theory
XI, 156 pages, 1987.

Vol. 88: Bruce A. Francis
A Course in H_∞ Control Theory
X, 150 pages, 1987.
Corrected - 1st printing 1987

Vol. 89: G. K. H. Pang/A. G. J. McFarlane
An Expert System Approach to Computer-Aided Design of Multivariable Systems
XII, 223 pages, 1987.

Vol. 90: Singular Perturbations and Asymptotic Analysis in Control Systems
Edited by P. Kokotovic, A. Bensoussan, G. Blankenship
VI, 419 pages, 1987.

Vol. 91 Stochastic Modelling and Filtering
Proceedings of the IFIP-WG 7/1 Working Conference
Rome, Italy, Decembre 10-14, 1984
Edited by A. Germani
IV, 209 pages, 1987.

Vol. 92: L. T. Grujić, A. A. Martynyuk, M. Ribbens-Pavella
Large-Scale Systems Stability Under Structural and Singular Perturbations
XV, 366 pages, 1987.

Vol. 93: K. Malanowski
Stability of Solutions to Convex Problems of Optimization
IX, 137 pages, 1987.

Vol. 94: H. Krishna
Computational Complexity of Bilinear Forms
Algebraic Coding Theory and Applications to Digital Communication Systems
XVIII, 166 pages, 1987.

Vol. 95: Optimal Control
Proceedings of the Conference on Optimal Control and Variational Calculus
Oberwolfach, West-Germany, June 15-21, 1986
Edited by R. Bulirsch, A. Miele, J. Stoer and K. H. Well
XII, 321 pages, 1987.

Vol. 96: H. J. Engelbert/W. Schmidt
Stochastic Differential Systems
Proceedings of the IFIP-WG 7/1 Working Conference
Eisenach, GDR, April 6-13, 1986
XII, 381 pages, 1987.

For information about Vols. 1–80 please contact your bookseller or Springer-Verlag.

Lecture Notes in Control and Information Sciences

Edited by M. Thoma and A. Wyner

IFIP

154

A. Kurzhanski,
I. Lasiecka (Eds.)

Modelling and Inverse Problems of Control for Distributed Parameter Systems

Proceedings of IFIP (W. G. 7.2)-IIASA Conference, Laxenburg, Austria, July 24-28, 1989

Springer-Verlag Berlin Heidelberg GmbH

Editor of Conference Proceedings of the series:
Computational Techniques in Distributed Systems IFIP-WG 7.2
Irena Lasiecka
Department of Applied Mathematics
Thornton Hall
University of Virginia
Charlottesville, VA 22903
USA

Editors
Professor Alexander Kurzhanski
System and Decision Sciences Programm
International Institute for Applied Systems Analysis
A-2361 Laxenburg
Austria

Professor Irena Lasiecka
Department of Applied Mathematics
University of Virginia
Charlottesville, VA 22903
USA

ISBN 978-3-540-53583-6 ISBN 978-3-540-46839-4 (eBook)
DOI 10.1007/978-3-540-46839-4

Originally published by Springer-Verlag London Berlin Heidelberg in 1991

Offsetprinting: Mercedes-Druck, Berlin

61/3020-543210 Printed on acid-free paper

Preface

The techniques of solving inverse problems that arise in the estimation and control of distributed parameter systems in the face of uncertainty as well as the applications of these to mathematical modelling for problems of applied system analysis (environmental issues, technological processes, biomathematical models, mathematical economy and other fields) are among the major topics of research at the Dynamic Systems Project of the System and Decision Sciences (SDS) Program at IIASA, the International Institute for Applied Systems Analysis.

In July 1989 the SDS Program was a coorganizer of a regular IFIP (WG 7.2) conference on Modelling and Inverse Problems of Control for Distributed Parameter Systems that was held at IIASA (Laxenburg, Austria), and was attended by a number of prominent theorists and practitioners. One of the main purposes of this meeting was to review recent developments and perspectives in this field. The proceedings are presented in this volume.

We believe that this conference has also achieved one of the goals of IIASA which is to promote and encourage cooperation between the scientists of East and West.

We wish to thank the Directorate and the Staff of IIASA for their contribution to the organization and the success of the conference. Our thanks goes particularly to Dr. A. Khapalov for his efforts in preparing this volume for publication.

Alexander B. Kurzhanski
System and Decision Sciences Program
International Institute for Applied Systems Analysis
Laxenburg, Austria

Irena Lasiecka
Department of Applied Mathematics
University of Virginia
Charlottesville, Virginia, USA

List of Participants

IFIP (WG. 7.2) – IIASA Conference

Modelling and Inverse Problems of Control for Distributed Parameter Systems

IIASA, Laxenburg, Austria

24–28 July, 1989

Professor P. Brunovsky
Institute of Applied Mathematics
Komensky University
Mlynska Dolina
Bratislava 84215
C.S.S.R.

Professor A. Brunovska
Institute of Applied Mathematics
Komensky University
Mlynska Dolina
Bratislava 84215
C.S.S.R.

Professor F. Colonius
University Bremen
Bibliothekstrasse
D-2800 Bremen, 33
F.R.G.

Professor J. Dolezal
Institute of Information Theory
and Automation
Czechoslovak Academy of Sciences
Pod vodarenskou vezi 4
18208 Prague 8
Czechoslovakia

Professor H.O. Fattorini
Department of Mathematics
University of California
Los Angeles, Ca 90024
U.S.A.

Professor Y. Galchuk
Moscow State University
Moscow
U.S.S.R.

Professor V.I. Gurman
Irkutsk Computer Center
ul. Lermontova 134
664033 Irkutsk 33
U.S.S.R.

Professor F. Kappel
Institute of Mathematics
Karl-Franzens-Universitaet Graz
Elisabethstrasse 11
A-8010 Graz
Austria

Professor R. Kazakova
Institute of Applied Mathematics
Miusskaya pl. 4
125047 Moscow
U.S.S.R.

Professor V.B. Kolmanovski
Moscow Institute of Electronic Machines
Dept. of Cybernetics
Bolshoi Vuzovskii 3/12
109028 Moscow
U.S.S.R.

Professor K. Kunisch
Institute of Mathematics
Technical University Graz
Kopernikusgasse 24
A-8010 Graz
Austria

Professor Irena Lasiecka
Department of Applied Mathematics
University of Virginia
Thorton Hall
Charlottesville, Virginia 22903
U.S.A.

Professor A. Lindquist
Department of Mathematics
Royal Institute of Technology
Stockholm
Sweden

Professor C. Martin
Chairman
Dept. of Mathematics
Texas Technical University
Lubbock, Texas, 79413
U.S.A.

Academician Yu. S. Osipov
Institute of Mathematics and Mechanics
of the Ural Scientific Center
Academy of Sciences of the USSR
Sverdlovsk

Dr. A.V. Pokatayev
Institute of Mathematics
Belorussian Academy of Sciences
Surganov Str. II
Minsk, 220604
U.S.S.R.

Professor G. Da Prato
Scuola Normale Superiore
I-56100 Pisa
Italy

Professor V.G. Romanov
Presidium of the Siberian Branch
of the USSR Academy of Sciences
Prospekt Acad. Lavrentjeva, 17
Novosibirsk
U.S.S.R.

Professor Tomas Roubicek
Institute of Information Theory & Automation
Czech. Acad. of Sciences
Pod vodarenskou vezi 4
182 08 Prague 8
C.S.S.R.

Academician A.A. Samarskii
Institute of Applied Mathematics
Miusskaya pl. 4
125047 Moscow
U.S.S.R.

Dipl. Ing. Otmar Scherzer
Institute für Mathematik
Univ. Linz
A-4040 Linz
Austria

Professor J. Sokolowski
Systems Research Institut, PAN
Control Theory Division
ul. Newelska 6,
Pl-01447 Warsaw
Poland

Professor V.A. Troitskii
LPI
ul. Politechnicheskaya, 29
195251 Leningrad
U.S.S.R.

Professor M.I. Zelikin
Mech.-Math. Faculty
Moscow State University
Leninskie Gory
Moscow
U.S.S.R.

Participants from IIASA's System and Decision Sciences Program

A. Kurzhanski (Chairman, USSR)
J.-P. Aubin (France)
H. Frankowska (France)
A. Gombani (Italy)
A. Khapalov (USSR)
M. Tanake (Japan)
V. Veliov (Bulgaria)

Contents

Tracking Property: a Viability Approach

Jean-Pierre Aubin

CEREMADE, Université de Paris-Dauphine
F-75775, Paris cx(16) France &
IIASA, International Institute for Applied Systems Analysis

Abstract

This paper is devoted to the characterization of the tracking property connecting solutions to two differential inclusions or control systems through an observation map derived from the viability theorem. The tracking property holds true if and only if the dynamics of the two systems and the contingent derivative of the observation map satisfy a generalized oartial differential equation, called the ***contingent differential inclusion***. This contingent differential inclusion is then used in several ways. For instance, knowing the dynamics of the two systems, construct the observation map or, knowing the dynamics of one system and the observation map, derive dynamics of the other system (trackers) which are solutions to the contingent differential inclusion.

It is also shown that the tracking problem provides a natural framework to treat issues such as the zero dynamics, decentralization, and hierarchical decomposition.

Introduction

Consider two finite dimensional vector-spaces X and Y, two set-valued maps $F : X \times Y \rightsquigarrow X$, $G : X \times Y \rightsquigarrow Y$ and the ***system of differential inclusions***

$$\begin{cases} x'(t) \in F(x(t), y(t)) \\ y'(t) \in G(x(t), y(t)) \end{cases}$$

We further introduce a set-valued map $H : X \rightsquigarrow Y$, regarded as an ***observation map***.

We devote this paper to many issues related to the following ***tracking property***: for every $x_0 \in \mathrm{Dom}(H)$ and every $y_0 \in H(x_0)$, there exist solutions $(x(\cdot), y(\cdot))$ to the system of differential inclusions such that

$$\forall\, t \geq 0, \quad y(t) \in H(x(t))$$

The answer to this question is a solution to a ***viability problem***, since we actually look for a solution $(x(\cdot), y(\cdot))$ which remains viable in the graph of the observation map H. So, if the set-valued maps F and G are Peano[1] maps and if the graph of H is closed, the Viability Theorem states that the tracking property is equivalent to the fact that the graph of H is a viability domain of $(x, y) \rightsquigarrow F(x, y) \times G(x, y)$.

Recalling that the graph of the contingent derivative $DH(x, y)$ of H at a point (x, y) of its graph is the contingent cone[2] to the graph of H at (x, y), the tracking property is then equivalent to the ***contingent differential inclusion***

$$\forall\, (x, y) \in \mathrm{Graph}(H), \quad G(x, y) \cap DH(x, y)(F(x, y)) \neq \emptyset$$

---

[1] A set-valued map is called ***Peano*** if its graph is nonempty and closed, its values are convex and its growth linear.

[2] The contingent cone $T_K(x)$ to a subset K at $x \in K$ is the closed cone of directions $v \in X$ such that $\lim_{h \to 0+} d_K(x + hv)/h = 0$. It is equal to X when x belongs to the interior of K, coincides with the tangent space when K is smooth and to the tangent cone of convex analysis when K is convex. We say that K if *sleek* at x is $y \rightsquigarrow T_K(y)$ is lower semicontinuous at x. In this case, the contingent cone $T_K(x)$ is convex. Convex subsets are sleek.

If (x, y) belongs to the graph of a set-valued map $H : X \rightsquigarrow Y$, the ***contingent derivative*** $DH(x, y)$ of H at (x, y) is the set-valued map from X to Y defined by

$$\mathrm{Graph}(DH(x, y)) := T_{\mathrm{Graph}(H)}(x, y)$$

We observe that when F and G are single-valued maps f and g and H is a differentiable single-valued map h, the contingent differential inclusion boils down to the more familiar ***system of first-order partial differential equations***[3]

$$\forall\, j = 1,\ldots,m,\ \sum_{i=1}^{n} \frac{\partial h_j}{\partial x_i} f_i(x,h(x)) - g_j(x,h(x)) \;=\; 0$$

Since the contingent differential inclusion links the three data F, G and H, we can use it in three different ways:

1. — Knowing F and H, find G or selections g of G such that the tracking property holds (observation problem)

2. — Knowing G (regarded as an ***exosystem***, following Byrnes-Isidori's terminology) and H, find F or selections of f of F such that the tracking property holds (tracking problem)

3. — Knowing F and G, find observation maps H satisfying the tracking property, i.e., solve the above contingent differential inclusion.

Furthermore, we can address other questions such as:

a) — Find the largest solution to the contingent differential inclusion (which then, contains all the other ones if any)

b) — Find single-valued solutions h to the contingent differential inclusion which then becomes

$$\forall\, x \in K,\ \ 0 \in Dh(x)(F(x,h(x))) - G(x,h(x))$$

In this case, the tracking property states that there exists a solution to the *"reduced" differential inclusion*

$$x'(t) \;\in\; F(x(t),h(x(t)))$$

so that $(x(\cdot),y(\cdot) := h(x(\cdot)))$ is a solution to the initial system of differential inclusions starting at $(x_0,h(x_0))$. Knowing h allows to divide the system by half, so to speak.

The observation and the tracking problems are the two sides of the same coin because the set-valued map H and its inverse play the same roles whenever we regard a single-valued map as a set-valued map characterized by its graph.

Consider then the observation problem: the idea is to observe solutions of a system $x' \in F(x,y)$ by a system $y' \in G(x,y)$ where $G : Y \rightsquigarrow Y$ describes simpler dynamics: equilibria, uniform movement, exponential growth, periodic solutions, etc. This would allow to observe complex systems[4] $x' \in F(x)$ in high dimensional spaces X by simpler systems $y' \in G(y)$ or even better, $y' = g(y)$, in low dimension spaces. We can think of H as an observation map, made of a small number of ***sensors*** taking into account uncertainty or lack of precision.

For instance, when $G \equiv 0$, we obtain constant observations. Observation maps H such that $F(x) \cap DH(x,y)^{-1}(0) \neq \emptyset$ for all $y \in H(x)$ provide solutions satisfying

$$\forall\, t \geq 0,\ \ x(t) \in H^{-1}(y_0) \ \text{ where } y_0 \in H(x_0)$$

In other words, inverse images $H^{-1}(y_0)$ are closed viability domains[5] of F. *Viewed through such an observation map, the system appears in equilibrium.*

---

[3]For special types of systems of differential equations, the graph of such a map h (satisfying additional properties) is called a ***center manifold***. Theorems providing the existence of local center manifolds have been widely used for the study of stability near an equilibrium and in control theory.

[4]We can use this tracking property as a ***mathematical metaphor*** to model the concept of .... metaphors in epistemology. The simpler system (the model) $y' \in G(y)$ is designed to provide ***explanations*** of the evolution of the unknown system $x' \in F(x)$ and the tracking property means that the ***metaphor*** H is valid (***non falsifiable***). Evolution of knowledge amounts to "increase" the observation space Y and to ***modify*** the system G (replace the model) and/or the observation map H (obtain more experimental data), checking that the tracking property (the validity or the consistency of the metaphor) is maintained.

[5]When $Y := \mathbf{R}$, such maps can be called "prime integrals" (or "energy functions") of F, because when both $F := f$ and $H := h$ are single-valued, we find the usual condition $h'(x) \cdot f(x) = 0$.

More generally, if there exists a linear operator $A \in \mathcal{L}(Y,Y)$ such that

$$\forall\, y \in \operatorname{Im}(H),\ \forall\, x \in H^{-1}(y),\quad F(x) \cap DH(x,y)^{-1}(Ay) \neq \emptyset$$

then we obtain solutions $x(\cdot)$ satisfying the time-dependent viability condition

$$\forall\, t \geq 0,\quad x(t) \in H^{-1}(e^{At}y_0) \text{ where } y_0 \in H(x_0)$$

so that we can use the exhaustive knowledge of linear differential equations to derive behavioral properties of the solutions to the original system.

But instead of checking whether such or such dynamics G satisfy the tracking property, we can look for systematic ways of finding them. For that purpose, it is natural to appeal to the selection procedures studied in [8, Chapter 6]. For instance, the most attractive idea is to choose the minimal selection $(x,y) \mapsto g^\circ(x,y)$ of the set-valued map

$$(x,y) \rightsquigarrow DH(x,y)(F(x,y))$$

which, by construction, satisfies the contingent differential inclusion. We shall prove that under adequate assumptions, the system

$$\begin{cases} i) & x'(t) \in F(x(t),y(t)) \\ ii) & y'(t) = g^\circ(x(t),y(t)) \end{cases}$$

has solutions (satisfying automatically the tracking property) even though the minimal selection g° is not necessarily continuous (see [15,4] for the use of minimal selections).

The drawback of the minimal selection and the other ones of the same family is that g° depends upon x. We would like to obtain single-valued dynamics g independent of x. They are selections of the set-valued map G_H defined by

$$G_H(y) := \bigcap_{x \in H^{-1}(y)} DH(x,y)(F(x,y))$$

We must appeal to Michael's Continuous Selection Theorem to find continuous selections g of this map, so that the system

$$\begin{cases} i) & x'(t) \in F(x(t),y(t)) \\ ii) & y'(t) = g(y(t)) \end{cases}$$

has solutions satisfying the tracking property.

The size of the set-valued map G_H *measures in some sense a degree of inadequacy of the observation of the system* $x' \in F(x)$ through H, because the larger the images of G_H, the more dynamics g tracking an evolution of the differential inclusion.

Tracking problems are intimately related to the observation problem: Here, the system $y' \in G(y)$, called the *exosystem*, is given, and so are their solutions when the initial states are fixed. The problem is to *regulate the system* $x'(t) \in F(x(t),y(t))$ for finding solutions $x(\cdot)$ *that match the solutions to the exosystem* $y'(t) \in G(y(t))$ in the sense that $y(t) \in H(x(t))$, or, more to the point, $x(t) \in H^{-1}(y(t))$.

Decentralization of control systems and *decoupling properties* are instances of this problem.

An instance of decentralization can be described as follows: We take $X := Y^n$, $F(x) := \prod_{i=1}^n F_i(x_i)$, and the viability subset is given in the form

$$K := \{(x_1,\ldots,x_n) \mid \sum_{i=1}^n x_i \in M\}$$

so that we observe the individual evolutions $x_i'(t) \in F_i(x_i(t))$ through their sum $y(t) := \sum_{i=1}^n x_i(t)$. Decentralizing the system means solving

— first a differential inclusion $y'(t) \in G(y(t))$ providing a solution $y(\cdot)$ viable in the viability subset $M \subset Y$, and

— second, find solutions to the differential inclusions $x_i'(t) \in F_i(x_i(t))$ satisfying the (time-dependent) viability condition

$$\sum_{i=1}^{n} x_i(t) = y(t)$$

condition which does not depend anymore on M.

Hierarchical decomposition happens whenever the observation map is a composition product of several maps determining the *successive levels of the hierarchy*. The evolution at each level is linked to the state of the lower level and is regulated by controls depending upon the evolution of the state-control of the lower level.

1 The Tracking Property

1.1 Characterization of the Tracking Property

Consider two finite dimensional vector-spaces X and Y, two set-valued maps $F : X \times Y \rightsquigarrow X$, $G : X \times Y \rightsquigarrow Y$ and a set-valued map $H : X \rightsquigarrow Y$, called the *observation map*:

Definition 1.1 *We shall say that F, G and H satisfy the* tracking property *if for any initial state $(x_0, y_0) \in \mathrm{Graph}(H)$, there exists at least one solution $(x(\cdot), y(\cdot))$ to the* system of differential inclusions

$$\text{(1)} \qquad \begin{cases} x'(t) & \in & F(x(t), y(t)) \\ y'(t) & \in & G(x(t), y(t)) \end{cases}$$

satisfying

$$\forall t \geq 0, \quad y(t) \; \in \; H(x(t))$$

We shall say that a set-valued map $H : X \rightsquigarrow Y$ is a solution to the contingent differential inclusion *if its graph is a* closed *subset of* $\mathrm{Dom}(F) \cap \mathrm{Dom}(G)$ *and if*

$$\text{(2)} \qquad \forall (x, y) \in \mathrm{Graph}(H), \quad G(x, y) \cap DH(x, y)(F(x, y))$$

We deduce at once from the viability theorems of [8, Chapter 3] the following:

Theorem 1.2 *Let us assume that $F : X \times Y \rightsquigarrow X$, $G : X \times Y \rightsquigarrow Y$ are Peano maps and that the graph of the set-valued map H is a closed subset of* $\mathrm{Dom}(F) \cap \mathrm{Dom}(G)$.

1. — The triple (F, G, H) enjoys the tracking property if and only if H is a solution to the contingent differential inclusion *(2).*

2. — There exists a largest *solution $H_\star$ to the contingent differential inclusion (2) contained in H. It enjoys the following property: whenever an initial state $y_0 \in H(x_0)$ does not belong to $H_\star(x_0)$, then all solutions $(x(\cdot), y(\cdot))$ to the system of differential inclusions (1) satisfy*

$$\text{(3)} \qquad \begin{cases} i) & \forall t \geq 0, \quad y(t) \notin H_\star(x(t)) \\ & \\ ii) & \exists T > 0 \quad \text{such that } y(T) \notin H(x(T)) \end{cases}$$

3. — If the set-valued maps $H_n \subset H$ are solutions to the contingent differential inclusion (2), so is their graphical upper limit[6].

[6]The *graphical upper limit* of a sequence of set-valued maps H_n is the set-valued map whose graph is the (Kuratowski) upper limit of the graphs of the H_n's.

We shall be interested in particular by single-valued solutions h to the partial contingent differential inclusion

$$\forall x \in K, \quad 0 \in Dh(x)(F(x,h(x))) - G(x,h(x))$$

In this case, the stability property implies the following statement: *Let us consider an equicontinuous sequence of continuous solutions h_n to the contingent differential inclusion converging pointwise to a function h. Then h is still a solution to the contingent differential inclusion.*

First, a pointwise limit h of single-valued maps h_n is a selection of the graphical upper limit of the h_n. The latter is equal to h when h_n remain in an equicontinuous subset: Indeed, in this case, any limit of elements $(x_n, h_n(x_n))$ being of the form $(x, h(x))$ belongs to the graph of h.

Remark — We could also introduce two other kinds of *contingent differential inclusions*:

$$\forall (x,y) \in \mathrm{Graph}(H), \quad DH(x,y)(F(x,y)) \subset G(x,y)$$

and

$$\forall (x,y) \in \mathrm{Graph}(H), \quad G(x,y) \subset \bigcap_{u \in F(x,y)} DH(x,y)(u)$$

The first inclusion implies obviously that any solution $(x(\cdot), y(\cdot))$ to the viability problem

$$x'(t) \in F(x(t), y(t)) \ \& \ x(t) \in H^{-1}(y(t))$$

parametrized by the absolutely continuous function $y(\cdot)$ is a solution to the differential inclusion

$$y'(t) \in G(x(t), y(t))$$

The second inclusion states that the graph of H is an invariance domain of the set-valued map $F \times G$. Assume that F and G are Lipschitz with compact values on a neighborhood of the graph of F. By the Invariance Theorem of [8, Theorem 5.4.5], the second inclusion is equivalent to the following strong tracking property:

For any initial state $(x_0, y_0) \in \mathrm{Graph}(H)$, *every solution* $(x(\cdot), y(\cdot))$ to the system of differential inclusions (1) starting at (x_0, y_0) satisfies $y(t) \in H(x(t))$ for all $t \geq 0$. □

We shall address now the problem of constructing *trackers*, which are selections of the set-valued map Φ

$$(x,y) \rightsquigarrow \Phi(x,y) := DH(x,y)(F(x,y))$$

For that purpose, we recall what we mean by selection procedure of a set-valued map F from a metric space X to a normed space Y.

1.2 Selection Procedures

Definition 1.3 (Selection Procedure) *Let X be a metric space, Y be a normed space and F be a set-valued map from X to Y. A* selection procedure *of a set-valued map $F : X \rightsquigarrow Y$ is a set-valued map $S_F : X \rightsquigarrow Y$ satisfying*

$$\begin{cases} i) & \forall x \in \mathrm{Dom}(F), \quad S(F(x)) := S_F(x) \cap F(x) \neq \emptyset \\ ii) & \text{the graph of } S_F \text{ is closed} \end{cases}$$

The set-valued map $S(F) : x \rightsquigarrow S(F(x))$ is called the selection *of F.*

The set-valued map defined by

$$(4) \qquad S_F^\circ(x,y) := \{v \in Y \mid \|v\| \leq d(0, F(x,y))\}$$

is naturally a selection procedure of a set-valued map with closed convex values which provides the minimal selection.

We can easily provide more examples of selection procedures through optimization thanks to the Maximum Theorem.

Proposition 1.4 *Let us assume that a set-valued map $F : X \rightsquigarrow Y$ is lower semicontinuous with compact values. Let V : Graph$(F) \mapsto \mathbf{R}$ be continuous. Then the set-valued map S_F defined by:*

$$S_F(x) := \{y \in Y \mid V(x,y) \leq \inf_{y' \in F(x)} V(x,y')\}$$

is a selection procedure *of F which yields selection $S(F)$ equal to:*

$$S(F(x)) = \{y \in F(x) \mid V(x,y) \leq \inf_{y' \in F(x)} V(x,y'))\}$$

Proof— Since F is lower semicontinuous, the function

$$(x,y) \mapsto V(x,y) + \sup_{y' \in F(x)} (-V(x,y'))$$

is lower semicontinuous thanks to the Maximum Theorem. Our proposition follows from :

$$\text{Graph}(S_F) = \{(x,y) \mid V(x,y) + \sup_{y' \in F(x)}(-V(x,y')) \leq 0\} \quad \square$$

Most selection procedures through game theoretical models or equilibria are instances of this general selection procedure based on Ky Fan's Inequality (see [3, Theorem 6.3.5] for instance).

Proposition 1.5 *Let us assume that a set-valued map $F : X \rightsquigarrow Y$ is lower semicontinuous with convex compact values. Let $\varphi : X \times Y \times Y \mapsto \mathbf{R}$ satisfy*

$$\begin{cases} i) & \varphi(x,y,y') \text{ is lower semicontinuous} \\ ii) & \forall (x,y) \in X \times Y, \ \ y' \mapsto \varphi(x,y,y') \text{ is concave} \\ iii) & \forall (x,y) \in X \times Y, \ \ \varphi(x,y,y) \leq 0 \end{cases}$$

Then the map S_F associated with φ by the relation

$$S_F(x) := \{y \in Y \mid \sup_{y' \in F(x)} \varphi(x,y,y') \leq 0\}$$

is a selection procedure of F yielding the selection map $x \mapsto S(F(x))$ defined by

$$S_F(x) := \{y \in F(x) \mid \sup_{y' \in F(x)} \varphi(x,y,y') \leq 0\}$$

Proof— Ky Fan's inequality states that the subsets $S_F(x)$ are not empty since the subsets $F(x)$ are convex and compact. The graph of S_F is closed thanks to the assumptions and the Maximum Theorem because it is equal to the lower section of a lower semicontinuous function:

$$\text{Graph}(S_F) = \{(x,y) \mid \sup_{y' \in F(x)} \varphi(x,y,y') \leq 0\} \quad \square$$

Proposition 1.6 *Assume that $Y = Y_1 \times Y_2$, that a set-valued map $F : X \rightsquigarrow Y$ is lower semicontinuous with convex compact values and that $a : X \times Y_1 \times Y_2 \to \mathbf{R}$ satisfies*

$$\begin{cases} i) & a \text{ is continuous} \\ ii) & \forall (x,y_2) \in X \times Y_2, \ \ y_1 \mapsto a(x,y_1,y_2) \text{ is convex} \\ iii) & \forall (x,y_1) \in X \times Y_1, \ \ y_2 \mapsto a(x,y_1,y_2) \text{ is concave} \end{cases}$$

Then the set-valued map S_F associating to any $x \in X$ the subset

$$\begin{array}{rcl} S_F(x) & := & \{(y_1,y_2) \in Y_1 \times Y_2 \text{ such that} \\ & & \forall (z_1,z_2) \in F(x), \ \ a(x,y_1,z_2) \leq a(x,z_1,y_2)\} \end{array}$$

is a selection procedure of F (with convex values). The selection map $S(F(\cdot))$ associates with any $x \in X$ the subset

$$\begin{array}{rcl} S(F)(x) & := & \{(y_1,y_2) \in F(x) \text{ such that} \\ & & \forall (z_1,z_2) \in F(x), \ \ a(x,y_1,z_2) \leq a(x,y_1,y_2) \leq a(x,z_1,y_2)\} \end{array}$$

of saddle-points *of $a(x,\cdot,\cdot)$ in $F(x)$.*

Proof — We take

$$\varphi(x,(y_1,y_2),(y_1',y_2')) := a(x,y_1,y_2') - a(x,y_1',y_2)$$

and we apply the above theorem. □

1.3 Construction of trackers

Any selection of the map Φ defined by

$$\forall\,(x,y) \in \mathrm{Graph}(H),\ \ \Phi(x,y) := DH(x,y)(F(x,y))$$

provides dynamics that satisfy the tracking property, provided that the system has solutions.

Naturally, we can obtain such selections by using selections procedures $G := S_\Phi$ of Φ (see Definition 1.3) that have convex values and linear growth, since the solutions to the system

$$\begin{cases} i) & x'(t) \in F(x(t),y(t)) \\ ii) & y'(t) \in S_\Phi(x(t),y(t)) \end{cases}$$

satisfying the tracking property (which exist by Theorem 1.2) are solutions to the system

$$\begin{cases} i) & x'(t) \in F(x(t),y(t)) \\ ii) & y'(t) \in S(\Phi)(x(t),y(t)) := \Phi(x(t),y(t)) \cap S_\Phi(x(t),y(t)) \end{cases}$$

Let us mention only the case of the minimal selection g° of Φ defined by

$$\begin{cases} i) & g^\circ(x,y) \in DH(x,y)(F(x,y)) \\ ii) & \|g^\circ(x,y)\| = \inf_{v\in DH(x,y)(F(x,y))} \|v\| \end{cases}$$

Theorem 1.7 *Assume that the Peano map F is continuous and that H is a sleek closed set-valued map satisfying, for some constant $c > 0$;*

$$\forall\,(x,y) \in \mathrm{Graph}(H),\ \ \|DH(x,y)\| \leq c$$

where $\|DH(x,y)\| := \sup_{\|u\|\leq 1} \inf_{v\in DH(x,y)(u)} \|v\|$ denotes the norm of the closed convex process $DH(x,y)$. Then the system observed by the minimal selection g° of $DH(\cdot,\cdot)(F(\cdot,\cdot))$

$$\begin{cases} i) & x'(t) \in F(x(t),y(t)) \\ ii) & y'(t) = g^\circ(x(t),y(t)) \end{cases}$$

has solutions enjoying the tracking property.

Proof — By [6, Theorem 3.1.1] ,the set-valued map $(x,y,u) \rightsquigarrow DH(x,y)(u)$ is lower semicontinuous. We deduce then from the lower semicontinuity of F that the set-valued map Φ is also lower semicontinuous. Since $DH(x,y)$ is a convex process, it maps the convex subset $F(x,y)$ to the convex subset $\Phi(x,y)$. Therefore, Φ being lower semicontinuous with closed convex images, its minimal selection S_Φ° defined by (4) is closed with convex values. Furthermore,

$$\|g^\circ(x,y)\| \leq c\|F(x,y)\| \leq c'(\|x\| + \|y\| + 1)$$

since $\|DH(x,y)\| \leq c$ and the growth of F is linear. Then the system

$$\begin{cases} i) & x'(t) \in F(x(t),y(t)) \\ ii) & y'(t) \in S_\Phi^\circ(x(t),y(t)) \cap c'(\|x(t)\| + \|y(t)\| + 1)B \end{cases}$$

has solutions enjoying the tracking property by Theorem 1.2. Therefore for almost all $t \geq 0$,

$$y'(t) \in \Phi(x(t),y(t)) \cap S_\Phi^\circ(x(t),y(t)) = g^\circ(x(t),y(t))\ \square$$

1.4 The Observation Problem

We consider the important case when $G : Y \rightsquigarrow Y$ does not depend upon x. Hence the contingent differential inclusion becomes

$$\forall\, x \in \mathrm{Dom}(H), \forall\; y \in H(x), \quad G(y) \cap DH(x,y)(F(x,y)) \neq \emptyset$$

Example Let us consider the case of *descriptor systems*

$$Ex'(t) \;=\; Ax(t) + Bu(t)$$

which we want to observe through $H \in \mathcal{L}(X,Y)$ by the linear equation

$$y'(t) = Gy(t)$$

where $G \in \mathcal{L}(Y,Y)$. We introduce the matrices (A, GH) from X to $X \times Y$ and

$$\begin{pmatrix} E & B \\ H & 0 \end{pmatrix} \text{ from } X \times Z \text{ to } X \times Y$$

We observe that the system enjoys the tracking property if and only if

$$\mathrm{Im}(A, GH) \subset \mathrm{Im} \begin{pmatrix} E & B \\ H & 0 \end{pmatrix}$$

In this case, the velocities $x'(t)$ and the controls $u(t)$ are supplied by the linear system

$$\begin{cases} Ex'(t) - Bu(t) & = Ax(t) \\ Hx'(t) & = GHx(t) \end{cases}$$

which can be solved by linear algebraic formulas. □

Example: Energy Maps (or Zero Dynamics) The simplest dynamics is obtained when $G \equiv 0$: in this case, each subset $H^{-1}(y)$ is a viability domain of $F(\cdot, y)$, because, for any $y \in \mathrm{Im}(H)$ and $x_0 \in H^{-1}(y)$, there exists a solution $x(\cdot)$ such that $x(t) \in H^{-1}(y_0)$ for all $t \geq 0$.

This viability property becomes:

$$\forall\, y \in \mathrm{Im}(H),\; \forall\, x \in H^{-1}(y), \quad F(x,y) \cap DH(x,y)^{-1}(0) \neq \emptyset$$

When F is a Peano map, we deduce that it is also equivalent to condition

$$\forall\, y \in \mathrm{Im}(H),\; \forall\, x \in H^{-1}(y), \quad F(x,y) \cap T_{H^{-1}(y)}(x) \;\neq\; \emptyset$$

We shall say that such a set-valued map H is an *energy map* of F.

In the general case, the evolution with respect to a parameter y of the viability kernels of the closed subsets $H^{-1}(y)$ under the set-valued map $F(\cdot, y)$ is described by the inverse of the largest solution $H_\star$:

Corollary 1.8 *Let $F : X \rightsquigarrow X$ be a Peano map. Then for any finite dimensional vector-space Y, there exists a largest closed energy map $H_\star : X \rightsquigarrow Y$ of F, a solution to the inclusion*

$$\forall\, x \in \mathrm{Dom}(H),\; \forall\, y \in H(x), \; DH(x,y)(F(x,y)) \ni 0$$

The inverse images $H_\star^{-1}(y)$ are the viability kernels of the subsets $H^{-1}(y)$ under the maps $F(\cdot, y)$:

$$\mathrm{Viab}_{F(\cdot,y)}(H^{-1}(y)) \;=\; H_\star^{-1}(y)$$

The graphical upper limit of energy maps is still an energy map.

Then *the graph of the map* $y \rightsquigarrow \text{Viab}_{F(\cdot,y)}(H^{-1}(y))$ *is closed, and thus upper semicontinuous whenever the domain of* H *is bounded.*

When the observation map H is a single-valued map h, the contingent differential inclusion becomes

$$\forall\, x,\ \exists\, u \in F(x,y) \ \text{ such that } \ 0 \in Dh(x)(u)$$

When h is differentiable and $F := f$ is single-valued, we find the classical characterization

$$< h'(x), f(x) > = \sum_{i=1}^{n} \frac{\partial h}{\partial x_i}(x) f_i(x) = 0$$

of *energy functions* or *prime integrals*[7] of the differential equation $x' = f(x)$.

The largest closed energy map contained in h is necessarily the restriction of h to a closed subset of the domain of h, *which is the viability kernel of* $h^{-1}(0)$. The restriction of the differential inclusion to the viability kernel of $h^{-1}(0)$ is (almost) what Byrnes and Isidori call the *zero dynamics of* F (in the framework of smooth nonlinear control systems).

Remark — The Equilibrium Map. We associate with each parameter y the set

$$E(y) \ := \ \{x \in H^{-1}(y) \mid 0 \in F(x,y)\}$$

of the equilibria of $F(\cdot, y)$ viable in $H^{-1}(y)$. We say that $E : Y \rightsquigarrow X$ is the *equilibrium map*.

We can derive some information on this equilibrium map from its derivative, which we can compute easily:

Theorem 1.9 *Assume that both* $H : X \rightsquigarrow Y$ *and* $F : X \times Y \rightsquigarrow X$ *are closed and sleek and that*

$$\begin{cases} \forall\, (x,y) \in \text{Graph}(H), \ \ \forall\, (u,v,w) \in X \times Y \times X, \\ \exists\, v_1 \in DH(x,y)(u_1) \ \ \textit{such that} \ w \in DF(x,y,0)(u+u_1, v+v_1) \end{cases}$$

Then the contingent derivative of the equilibrium map is the equilibrium map of the derivative:

$$u \in DE(y,x)(v) \Longleftrightarrow u \in DH(x,y)^{-1}(v) \ \ \& \ \ 0 \in DF(x,y,0)(u,v)$$

Proof — We observe that by setting $\pi(x,y) := (x,y,0)$, the graph of E^{-1} can be written:

$$\text{Graph}(E^{-1}) \ := \ \text{Graph}(H) \cap \pi^{-1}(\text{Graph}(F))$$

and we apply [6, Theorem 4.3.3], which states that if the transversality condition: for all $(x,y) \in \text{Graph}(E^{-1})$,

$$\pi\left(T_{\text{Graph}(H)}(x,y)\right) - T_{\text{Graph}(F)}(\pi(x,y)) \ = \ X \times Y \times X$$

holds true, then

$$T_{\text{Graph}(E^{-1})}(x,y) \ := \ T_{\text{Graph}(H)}(x,y) \cap \pi^{-1}\left(T_{\text{Graph}(F)}(\pi(x,y))\right)$$

Recalling that the contingent cone to the graph of a set-valued map is the graph of its contingent derivative, the assumption of our proposition implies the transversality condition. We then observe that the latter equality yields the conclusion of the proposition. □

Using the inverse function and the localization theorems presented in [6, section 5.4], we can derive the same kind of informations as the ones provided by [6, Proposition 5.4.7.].

For instance, set

$$Q(y,x) \ := \ u \in DH(x,y)^{-1}(0) \mid 0 \ \in \ DF(x,y,0)(u,0)$$

[7] When f is real-valued, this is the "contingent version" of the Hamilton-Jacobi equation. See the the papers and the forthcoming monograph by Frankowska [16] for an exhaustive study and the connections with the *viscosity solutions*.

Then, for any equilibrium $x \in E(y)$ and any closed cone P satisfying $P \cap Q(y,x) = \{0\}$, there exists $\varepsilon > 0$ such that

$$E(y) \cap (x + \varepsilon(P \cap B)) = \{x\}$$

where B denotes the unit ball. In particular, *an equilibrium $x \in E(y)$ is locally unique whenever $0 \in DH(x,y)^{-1}(0)$ is the unique equilibrium of $DF(x,y,0)(\cdot,0)$.*

Furthermore, if the set $E(y)$ of equilibria is convex, then

$$E(y) \subset x + Q(y,x) \quad \square$$

More generally, the behavior of observations of some solutions to the differential inclusion $x' \in F(x,y)$ may be given as the prescribed behavior of solutions to differential equations $y' = g(y)$, where g is a selection of

$$g(y) \in \bigcap_{x \in H^{-1}(y)} DH(x,y)(DF(x,y))$$

In the case when the differential equation $y' = g(y)$ has a unique solution $r(t)y_0$ staring from y_0, the solution $x(\cdot)$ satisfies the condition

$$\forall\, t \geq 0,\;\; x(t) \in H^{-1}(r(t)y(0)),\;\; x(0) \in H^{-1}(y(0))$$

When g is a linear operator $G \in \mathcal{L}(Y,Y)$, it can be written

$$\forall\, t \geq 0,\;\; x(t) \in H^{-1}(e^{Gt}y(0)),\;\; x(0) \in H^{-1}(y(0))$$

When $H \equiv h$ is a single-valued differentiable map, then the map G_H can be written

$$G_H(y) := \bigcap_{h(x)=y} h'(x)F(x,y)$$

and a single-valued map g is a selection of G_H if and only if

$$\forall\, x \in \mathrm{Dom}(H),\;\; 0 \in h'(x)F(x,y) - g(h(x))$$

The problem arises to construct such maps g.

1.5 Construction of Observers

These maps g are selections of the map $G_H : Y \leadsto Y$ defined by

$$G_H(y) := \bigcap_{x \in H^{-1}(y)} (DH(x,y)(F(x,y)))$$

(The set-valued map G_H *measures so to speak a degree of disorder of the system* $x' \in F(x,y)$, because the larger the images of G_H, the more observed dynamics g tracking an evolution of the differential inclusion.)

By using Michael's Continuous Selection Theorem, we obtain the following

Theorem 1.10 *Assume that the set-valued map F is continuous with convex compact images and linear growth, that H is a sleek closed set-valued map the domain of which is bounded and that there exists a constant $c > 0$ such that*

$$\forall\, (x,y) \in \mathrm{Graph}(H),\;\; \|DH(x,y)\| \leq c$$

Assume also that there exist constants $\delta > 0$ and $\gamma > 0$ such that, for any map $x \mapsto c(x) \in \gamma B$,

$$\delta B \cap \bigcap_{x \in H^{-1}(y)} (DH(x,y)(F(x,y)) - c(x)) \neq \emptyset$$

Then there exists a continuous map g such that the solutions of

$$\begin{cases} i) & x'(t) \in F(x(t), y(t)) \\ ii) & y'(t) = g(y(t)) \end{cases}$$

enjoy the tracking property for any initial state $(x_0, y_0) \in \text{Graph}(H)$.

Proof— The proof of the above theorem showed that the set-valued map Φ is lower semicontinuous with compact convex images. Furthermore, the set-valued map H^{-1} is upper semicontinuous with compact images since we assumed the domain of H bounded. Then the lower semicontinuity criterion [6, Theorem 1.5.3] implies that the set-valued map G_H is also lower semicontinuous with compact convex images. Then there exists a continuous selection g of G_H, so that the above system does have solutions viable in the graph of H. □

2 The Tracking Problem

2.1 Tracking Control Systems

Let $H : X \rightsquigarrow Y$ be an observation map. We consider two control systems

(5) $$\begin{cases} i) & \text{for almost all } t \geq 0, \; x'(t) = f(x(t), u(t)) \\ ii) & \text{where } u(t) \in U(x(t)) \end{cases}$$

and

(6) $$\begin{cases} i) & \text{for almost all } t \geq 0, \; y'(t) = g(y(t), v(t)) \\ ii) & \text{where } v(t) \in V(y(t)) \end{cases}$$

on the state and observation spaces respectively, where $U : X \rightsquigarrow Z_X$ and $V : Y \rightsquigarrow Z_X$ map X and Y to the control spaces Z_X and Z_Y and where $f : \text{Graph}(U) \mapsto X$ and $g : \text{Graph}(V) \mapsto Y$.

We introduce the set-valued maps $R_H(x, y) : Z_Y \rightsquigarrow Z_X$ defined by

$$R_H(x, y; v) = \begin{cases} \{u \in U(x) | f(x, u) \in DH(x, y)^{-1}(g(y, v))\} & \text{if } v \in V(y) \\ \emptyset & \text{if } v \notin V(y) \end{cases}$$

Corollary 2.1 *Assume that the set-valued maps U and V are Peano maps and that the maps f and g are continuous, affine with respect to the controls and with linear growth. The two control systems enjoy the tracking property if and only if*

$$\forall (x, y) \in \text{Graph}(H), \quad \text{Graph}(R_H(x, y)) \neq \emptyset$$

Then the system is regulated by the regulation law

$$\text{for almost all } t \geq 0, \quad u(t) \in R_H(x(t), y(t); v(t))$$

When $H \equiv h$ is single-valued and differentiable and when we set $f(x, u) := c(x) + g(x)u$ and $g(y, v) := d(y) + e(y)v$ where $g(x)\cdot$ and $e(y)\cdot$ are linear operators, we obtain the formula

$$R_h(x; v) := U(x) \cap (h'(x)g(x))^{-1}(d(h(x)) - h'(x)c(x) + e(h(x)v))$$

2.2 Decentralization of a control system

We assume that the viability set of the control system (5) is defined by constraints of the form $K := L \cap h^{-1}(M)$ where

(7) $$\begin{cases} i) & L \subset X \text{ and } M \subset Y \text{ are sleek} \\ ii) & h \text{ is a } C^1\text{-map from } X \text{ to } Y \\ iii) & \forall x \in K := L \cap h^{-1}(M), \; Y = h'(x)T_L(x) - T_M(h(x)) \end{cases}$$

We associate with the two systems (5), (6) the *decoupled viability constraints*

$$(8) \qquad \begin{cases} i) & \forall t \geq 0, \ x(t) \in L \\ ii) & \forall t \geq 0, \ h(x(t)) = y(t) \\ iii) & \forall t \geq 0, \ y(t) \in M \end{cases}$$

It is obvious that the *state component* $x(\cdot)$ of any solution $(x(\cdot), y(\cdot))$ to the system ((5),(6)) satisfying viability constraints (8) is a solution to the initial control system (5) viable in the set K defined by (7).

On the other hand, solutions to the system (5) viable in K can be obtained in two steps:

— first, find a solution $y(\cdot)$ to the control system (6) *viable in* M and then,

— second, find a solution $x(\cdot)$ the control system (5) satisfying the viability constraints

$$(9) \qquad \begin{cases} i) & \forall t \geq 0, \ x(t) \in L \\ ii) & \forall t \geq 0, \ h(x(t)) = y(t) \end{cases}$$

which no longer involve the subset $M \subset Y$ *of constraints.*

This decentralization problem is a particular case of the observation problem for the set-valued map H defined by

$$H(x) := \begin{cases} h(x) & \text{if } x \in L \ \& \ h(x) \in M \\ \emptyset & \text{if not} \end{cases}$$

whose contingent derivative is equal under assumptions (7) to

$$DH(x)(u) := \begin{cases} h'(x)u & \text{if } u \in T_L(x) \ \& \ h'(x)u \in T_M(h(x)) \\ \emptyset & \text{if not} \end{cases}$$

We know that the regulation map of the initial system (5), (6) on the subset K defined by (7) is equal to

$$R_K(x) = \{u \in U(x) \cap T_L(x) \mid h'(x)f(x,u) \in T_M(h(x))\}$$

The regulation map of the projected control system (6) on the subset M is defined by

$$R_M(y) = \{v \in V(y) \mid g(y,v) \in T_M(y)\}$$

We introduce now the set-valued map R_H which is equal to

$$R_H(x,y;v) := \{u \in U(x) \cap T_L(x) \mid h'(x)f(x,u) = g(y,v)\}$$

We observe that

$$\forall x \in K, \ R_H(x, h(x); R_M(h(x))) \subset R_K(x)$$

The regulation map regulating solutions to the system ((5),(6)) satisfying viability conditions (8) is equal to $x \rightsquigarrow R_H(x, h(x); R_M(h(x)))$. Therefore, the regulation law feeding back the controls from the solutions are given by: for almost all $t \geq 0$

$$\begin{cases} i) & v(t) \in R_M(y(t)) \\ ii) & u(t) \in R_H(x(t); v(t)) \end{cases}$$

The first law regulates the solutions to the control system (6) viable in M *and the second regulates the solutions to the control system (5) satisfying the viability constraints (9).*

Remark — The reason why this property is called decentralization lies in the particular case when $X := Y^n$, when $h(x) := \sum_{i=1}^n x_i$ and when the control system (5) is

$$\forall\, i = 1, \ldots, n, \ \ x_i'(t) = f_i(x_i(t), u_i(t)) \ \text{ where } u_i(t) \in U_i(x_i(t))$$

constrained by

$$\forall\, i = 1, \ldots, n, \quad x_i(t) \in L_i \;\; \& \;\; \sum_{i=1}^{n} x_i(t) \in M$$

We introduce the regulation map R_H defined by

$$\begin{array}{l} R_H(x_1, \ldots, x_n, y; v) \\ := \; \{u \in \bigcap_{i=1}^n (U_i(x_i) \cap T_{L_i}(x_i)) \mid \sum_{i=1}^n f_i(x_i, u) = g(y, v)\} \end{array}$$

This system can be decentralized first by solving the viability problem for system (6) in the viability set M through the regulation law $v(t) \in R_M(y(t))$.

This being done, the state-control $(y(\cdot), v(\cdot))$ being known, it remains in a second step to study the evolution of the n control systems

$$\forall\, i = 1, \ldots, n, \quad x_i'(t) \;=\; f_i(x_i, u(t))$$

through the regulation law

$$u(t) \;\in\; R_H(x_1(t), \ldots, x_n(t), y(t); v(t)) \;\; \square$$

Economic Interpretation — We can illustrate this problem with an economic interpretation: the state $x := (x_1, \ldots, x_n)$ describes an allocation of a commodity $y \in M$ among n consumers. The subsets L_i represent the consumptions sets of each consumer and the subset M the set of available commodities. The control u plays the role of the price system of the consumptions goods and v the price of the production goods. Differential equations $x_i' = f_i(x_i, u)$ represent the behavior of each consumer in terms of the consumption price and $y' = g(y, v)$ the evolution of the production process.

The decentralization process allows us to decouple the production problem and the consumption problem. See more details in [8, Chapter 15] on dynamical economic models. $\square$

2.3 Hierarchical Decomposition Property

For simplicity, we restrict ourself here to the case when *the observation map* $H \equiv h := h_2 \circ h_1$ *is the product of two differentiable single-valued maps* $h_1 : X \mapsto Y_1$ and $h_2 : Y_1 \mapsto Y_2$.

We address the following issue: Can we observe the evolution of a solution to a control problem (5) through $h_2 \circ h_1$ by observing it

— first through h_1 by a control system

(10)
$$\left\{ \begin{array}{ll} i) & \text{for almost all } t \geq 0, \;\; y_1'(t) \;=\; g_1(y_1(t), v_1(t)) \\ ii) & \text{where } \; v_1(t) \;\in\; V_1(y_1(t)) \end{array} \right.$$

and then,

— second, observing this system through h_2.

We introduce the maps R_h, R_{h_1} and R_{h_2} defined respectively by

$$\left\{ \begin{array}{ll} R_h(x; v) & := \{u \in U(x) \mid h'(x) f(x, u) = g(h(x), v) \\ & \quad \text{if } v \in V(h(x))\} \\ & \\ R_{h_1}(x; v_1) & = \{u \in U(x) \quad h_1'(x) f(x, u) = g_1(h_1(x), v_1) \\ & \quad \text{if } v_1 \in V(h_1(x))\} \\ & \\ R_{h_2}(x_1; v) & = \{v_1 \in V_1(x_1) \mid h_2'(x_1) g_1(x_1, v_1) = g(h_2(x_1), v) \\ & \quad \text{if } v \in V(h_2(x_1))\} \end{array} \right.$$

and we see at once that

$$R_{h_1}(x; R_{h_2}(h_1(x); v)) \;\subset\; R_h(x; v)$$

Therefore, if the graph of $v \rightsquigarrow R_{h_1}(x; R_{h_2}(h_1(x); v))$ is not empty, we can recover from the evolution of a solution $y(\cdot)$ to the control system (6) a solution $y_1(\cdot)$ to the control system (10) by the tracking law

$$\text{for almost all } t,\quad v_1(t) \in R_{h_2}(y_1(t), v(t))$$

and then, a solution $x(\cdot)$ to the control system (5) by the tracking law

$$\text{for almost all } t,\quad u(t) \in R_{h_1}(x(t), v_1(t))$$

This can illustrate hierarchical organization *which is found in the evolution of so many macro-systems. The decomposition of the observation map as a product of several maps determines the successive levels of the hierarchy. The evolution at each level obeys the constraint binding its state to the state of the lower level. It is regulated by controls determined (in a set-valued way) by the evolution of the state-control of the lower level.*

References

[1] AUBIN J.-P. , BYRNES C. & ISIDORI A. (1990) *Viability Kernels, Controlled Invariance and Zero Dynamics for Nonlinear Systems*, Proceedings of the 9th International Conference on Analysis and Optimization of Systems, Nice, June 1990, Lecture Notes in Control and Information Sciences, Springer-Verlag

[2] AUBIN J.-P. & CELLINA A. (1984) DIFFERENTIALINCLUSIONS, Springer-Verlag

[3] AUBIN J.-P. & EKELAND I. (1984) APPLIED NONLINEAR ANALYSIS, Wiley-Interscience

[4] AUBIN J.-P. & FRANKOWSKA H. (1984) *Trajectoires lourdes de systèmes contrôlés*, Comptes-rendus de l'Académie des Sciences, PARIS , Série 1, 298, 521-524

[5] AUBIN J.-P. & FRANKOWSKA H. (1985) *Heavy viable trajectories of controlled systems*, Annales de l'Institut Henri Poincaré, Analyse Non Linéaire, 2, 371-395

[6] AUBIN J.-P. & FRANKOWSKA H. (1990) Birkhäuser SET-VALUED ANALYSIS,

[7] AUBIN J.-P. (1990) *A Survey of Viability Theory*, SIAM J. on Control and Optimization, 28, 749-788

[8] AUBIN J.-P. (to appear) VIABILITY THEORY

[9] BYRNES C.I. & ANDERSON B.D.O. (1984) *Output feedback and generic stabilizability* , SIAM J. Control and Optimization, 22 (3), 362-379

[10] BYRNES C. & ISIDORI A. (to appear) *The Analysis and Design of Nonlinear Feddback Systems. I. Zero Dynamics and Global Normal Forms*,

[11] BYRNES C. & ISIDORI A. (to appear) *The Analysis and Design of Nonlinear Feddback Systems .II Global Stabilization of Minimum Phase Systems*,

[12] BYRNES C. & ISIDORI A. (to appear) *Feedback Design From the Zero Dynamics Point of View*,

[13] BYRNES C. & ISIDORI A. (to appear) *Output Regulation of Nonlinear Systems*,

[14] BYRNES C. & ISIDORI A. (this volumr)

[15] FALCONE M. & SAINT-PIERRE P. (1987) *Slow and quasi-slow solutions of differential inclusions*, J. Nonlinear Anal.,T.,M.,A., 3, 367-377

[16] FRANKOWSKA H. ((to appear)) SET-VALUED ANALYSIS AND CONTROL THEORY (MONOGRAPH),

[17] FRANKOWSKA H. ((to appear)) *Some inverse mapping theorems*,

[18] HADDAD G. (1981) *Monotone viable trajectories for functional differential inclusions*, J. Diff. Eq., 42, 1-24

[19] HADDAD G. (1981) *Monotone trajectories of differential inclusions with memory*, Israel J. Maths, 39, 38-100

[20] ISIDORI A. (1985) NONLINEAR CONTROL SYSTEMS: AN INTRODUCTION, Springer-Verlag Lecture Notes in Control and Information Sciences, Vol.72

[21] KRENER A. & ISIDORI A. (1980) *Nonlinear Zero Distributions*, 19th IEEE Conf. Decision and Control

[22] KRENER A. J. & ISIDORI A. (1983) *Linearization by output injection and nonlinear observers*, Syst. & Control letters, 3, 47-52

[23] KURZHANSKII A. B. & FILIPPOVA T. F. (1986) *On viable solutions for uncertain systems*,

[24] KURZHANSKII A. B. (1985) *On the analytical description of the viable solutions of a controlled system*, Uspekhi Mat. Nauk, 4

[25] KURZHANSKII A. B. (1986) *On the analytical properties of viability tubes of trajectories of differential systems*, Doklady Acad. Nauk SSSR, 287, 1047-1050

[26] MARRO G. (1975) FONDAMENTI DI TEORIA DEI SISTEMI, Patron Editore

[27] MONACO S. & NORMAND-CYROT D. (1988) *Zero Dynamics of Sampled Linear Systems*, Systems and Control Letters

Extensions and Global Estimates for Evolutionary Discrete Control Systems

V. Gurman
Program Systems Institute (PSI)
Pereslavl-Zalessky, USSR
and
International School for Advanced Studies (ISAS),
Trieste, Italy

Abstract

This paper deals with the general procedure of the extension principle for the abstract evolutionary control system (in the time-discrete form) and estimation of the system's reachable set as an important characteristic. This is a new stage of the development of an approach that initially had been expressed in the well-known Krotov sufficient optimality conditions and proved to be very fruitful in applications for the lumped-parameter systems control problems.

1 Introduction

Presently, the class of distributed parameter systems that are studied with mathematical tools extends very far. Difficulties and problems arise in modeling new complex objects (such as ecological scenarios) and in choosing the strategy to investigate practical problems and in the methods to interpret the results (openness, uncertainty, impossibility or high cost of the strict observations and their discrete nature). As a result the model concept turns to have alternative versions (continuous, discrete, chamber, linear, nonlinear, degree of its detailization, etc.).

Therefore adaptive mathematical tools should be developed. From this point of view the methods of extensions and global estimates that were initially developed within optimal control theory are of particular interest [Krotov and Gurman (1973) and Krotov (1988)]. In particular, some important ecological control models with distributed parameters that are singular and nontraditional for mathematical physics and mathematical biology have been successfully studied and interesting optimal solutions, such as optimal dynamical forest structure and optimal fish population control have been obtained (Moskalenko, 1983). Current accumulated experience of the application of those tools to a broad class of lumped-parameter systems [Gurman (1985) and Konstantinov (1983)] gives us hope to be able to apply them to distributed-parameter systems with a possibility of quantitative estimation of object behavior bounds depending on available information and computational recourses.

This paper gives additional reasons to consider the abstract evolutionary system in the time-discrete form. This form requires the fewest function-theoretic properties of all the constructions to be used and allows us to concentrate on the proposed scheme due to its full invariance with respect to spaces structure and continuity concepts. On the other hand, recent results in infinite-dimensional differential inclusions theory (Tolstonogov, 1986) allow to use the correct transition from the time continuous form to the time-discrete form of the same evolutionary system either precisely or approximately.

2 The Discrete Evolutionary Model and Its Extensions

Let us consider (as an arbitrary evolutionary system model) the following chain relation

$$x(t+1) = f(t, x(t), u(t)), t = \{t_i, t_i + 1, \ldots, t_f\}\ , \tag{1}$$

$$x(t) \in X(t) \subset X_O(t_i), u(t) \in U(t, x(t)) \subset U_O(t)\ ,$$

where t is a number of the time registration; $X_O(t)$, $U_O(t)$ are basic set spaces of arbitrary nature (may be different for different (t); $X(t)$, $U(t, x(t))$ are given subsets of $X_O(t)$, $U_O(t)$; $f(t,\cdot) : X_O(t) \times U_O(t) \rightarrow X_O(t+1)$ is some given operator; and $x(t)$, $u(t)$ stand for the state and the external influence (or control) description at time t.

Another form of (1) is

$$x(t+1) \in \Pi(t, x(t)) \qquad \text{(1a)}\ ,$$

where $\Pi(t, x(t)) = f(t, x(t), U(t, x(t)))$.

We introduce an arbitrary family of maps $(\varphi_\alpha(t,\cdot) : X_O(t) \rightarrow Y_O(t))_\alpha$, $\alpha \in A$, and the corresponding family of new evolutionary systems (the derived systems)

$$y_\alpha(t+1) = \varphi_\alpha(t+1, f(t, x(t), u(t))), u(t) \in U(t, x(t)), x(t) \in Q(t, y(t)) = \varphi^{-1}(t, y(t))\ . \tag{2}$$

Another form of the derived system:

$$y_\alpha(t+1) \in \bigcup_{x(t) \in \phi^{-1}(t, y(t))} \varphi_\alpha(t+1, \Pi(t, x(t)))\ . \tag{2a}$$

Any of these systems is an extension of the initial system (1) in the sense that any solution to the initial system is also a solution to the derived system (but not vice-versa in the general case).

The estimates of such important characteristics of the system (1) as optimal regimes, attainability and controllability sets and related questions, may be obtained this way through the proper choice of φ_α.

3 Estimation of Reachable Set

In the general case some constructive procedures for the estimation of the reachable set $X_r(t)$ can be developed under $Y_{O\alpha} = R$, $\varphi_\alpha(t_i) : X_O(t) \to R$. The following auxiliary system is considered:

$$z_\alpha(t+1) = h_\alpha(t, z(t)) \geq \sup_{u \in U(t,x(t))} \varphi_\alpha(t+1, f(t, x(t), u(t))) \tag{3}$$

$$x(t) \in \Lambda_\alpha(t,z) = \bigcap_{\beta \in A} \{x(t) : \varphi_\alpha(t, x(t)) = z_\alpha(t), \varphi_\beta(t, x(t)) \leq z_\beta(t), \beta \neq \alpha\} \bigcap M(t),$$

$$z_\alpha(t_i) = \sup \varphi_\alpha(t_i, X_i) \,,$$

where $M(t)$ is some a priori external estimate for $X_r(\cdot)$, $M(t) \supset X_r(t)$ in particular, the trivial one, $X_O(t)$; z and designates the whole family of z_α, $\alpha \in A$. Denote

$$x_\varphi(t_i, X_i, t) = \bigcap_{\alpha \in A} \{x(t) : \varphi_\alpha(t, x(t)) \leq \hat{z}_\alpha(t)\} \bigcap M(t) \,,$$

where $\hat{z}_\alpha(t)$ is the solution to (3).

Theorem 3.1 *Any system (φ, h, M) (where φ, h designate the whole families of φ_α, h_α, $\alpha \in A$) defines an external estimate of the reachable set of the system for each $t \in T : X_\varphi(t) \supset X_r(t)$.*

Proof For $t = t_i$ we have $x(t_i) \in X_r(t_i) = X_i$. Hence,

$$x(t_i) \in X_\varphi(t_i) = \bigcap_{\alpha \in A} \{x(t_i) : \varphi_\alpha(t_i, x(t_i)) \leq \hat{z}_\alpha(t_i)\} \bigcap M(t_i)$$

so $X_r(t_i) \subset X_\varphi(t_i)$. Now let us show that

$$X_r(t) \subset X_\varphi(t) \Rightarrow X_\varphi(t+1) \supset X_r(t+1) \,.$$

The left-hand side of this implication means that for any $x(t) \in X_r(t)$

$$\varphi_\alpha(t, x(t)) = y_\alpha(t) \leq \hat{z}_\alpha(t) \quad \forall \alpha \in A \,. \tag{$*$}$$

Then we observe that $\Lambda_\alpha(t,z)$ can be represented by

$$\Lambda_\alpha(t,z) = \{x(t) : \varphi_\alpha(t, x(t)) = \hat{z}_\alpha(t)\} \bigcap M(t) \bigcap X_\varphi(t) \,.$$

Taking into account this representation, with $(*)$ and (3) we obtain $y_\alpha(t+1) = \varphi_\alpha(t+1, x(t+1)) \leq \hat{z}_\alpha(t+1)$ $\forall\, \alpha \in A$ when $x(t+1) \in X_r(t+1)$. Hence,

$$X_\varphi(t+1) = \bigcap_{\alpha \in A} \{x(t) : \varphi_\alpha(t+1, x(t+1)) \leq \hat{z}_\alpha(t+1)\} \bigcap M(t) \ni x(t+1) \,,$$

i.e., $X_r(t+1) \subset X_\varphi(t+1)$. This is the basis of mathematical induction reasoning that accomplishs the proof. It is important that this class of estimates contains (under some additional assumptions) the exact one that coincides with the reachable set. □

Theorem 3.2 *Let X_i be described by the inequality*

$$X_i = \{x(t_i) : \kappa(x(t_i)) \le q, x : X_O(t_i) \to R\} ,$$

$U_*(t, x(t)) = \operatorname*{Arg\,sup}_{u(t)\in U(t,x(t))} \varphi(t+1, f(t, x(t), u(t))) \ne 0$ *and a map* $\varphi(t,\cdot) : X_O(t) \to R$ *satisfies the following conditions*

$$\sup\nolimits_{u\in U(t,x(t))}(\varphi(t+1, f(t, x(t), u(t))) = c(t, \varphi(t, x(t))) , \tag{4}$$

$\varphi(t_i, x(t_i)) = \kappa(t_i, x(t_i))$ *with an arbitrary continuous monotone function* $c(t,\cdot) : R \to R$. *Then*

$$X_\varphi(t) \triangleq \{x(t) : \varphi(t, x(t)) \le z(t)\} = X_r(t) ,$$

where $z(t)$ is the solution of the chain

$$z(t+1) = c(t, z(t)), z(t_i) = q . \tag{5}$$

Proof It is clear that $X_\varphi(\cdot) \supset X_r(\cdot)$. Let us show that $X_\varphi(\cdot) \subset X(\cdot)$, considering the subsystem

$$x(t+1) \in f(t, x(t), U_*(t, x(t))) , \tag{6}$$

$x(t_i) \in X_i$, where the reachable set X_r^* is contained in $X_r(\cdot)$, because of $U_*(\cdot) \subset U(\cdot)$. Any solution image of this system $z(t) = \varphi(t, x(t))$ satisfies (5) due to condition (4). Take any element $x_\varphi(\tau) \in X_\varphi(\tau)$, $\tau \in T$ as an initial one for (6) and solve this chain from right to left to receive $x_\varphi(t)$, $t \in \{t_i, \ldots, \tau\}$ and the corresponding image $z_\varphi(t)$. Since $z_\varphi(\tau) = \varphi(\tau, x_\varphi(\tau)) \le z(\tau)$ then $z_\varphi(t) \le z(t)$ for each t due to the monotonicity of $c(t,\cdot)$. Hence, $z_\varphi(t_i) \le z(t_i) \le q$, i.e., $x_\varphi(t_i) \in X_i$ and $x_\varphi(\tau) \in X_r(\tau)$. This means that $X_\varphi(\cdot) \subset X_r$. □

Corollary 3.3 *If φ satisfies all the conditions of the theorem, excluding $X_i = \{x(t_i) : x(x(t_i)) \le q\}$ and $\varphi(t_i, x(t_i)) = \kappa(x(t_i))$ and if*

$$\max \varphi(t_i, X_i) = \varphi(t_i, x_*(t_i)) = q ,$$

then the set $X_\varphi = \{\varphi(t, x(t)) \le z(t)\}$ is an external estimate of $X_r(\cdot)$ for any $t \in T$ and $\varphi(t, x_*(t)) = z(t)$, where $x_*(t)$, $z(t)$ are solutions of equations (5) and (6). In other words the "bound" of $X_\varphi(\cdot)$ touches $X_r(\cdot)$ at any $t \in T$.

Example 1. Consider the following system in normed linear space

$$x(t+1) = \lambda x(t) + u(t), \|u\| \le 1, x(0) = 0 = \{x : \|x\| = 0\}, \lambda \in R .$$

Take $\varphi = \|x\|$. Then

$$y(t+1) = \|\lambda x(t) + u(t)\|$$

$$z(t+1) = \sup_{u,x} \|\lambda x(t) + u(t)\| = \sup_{x} \lambda \|x(t)\| + 1 = \lambda z(t) + 1 .$$

Therefore the map φ satisfies (4) under $c(\cdot) = \lambda z + 1$. We will find the solution of the last chain when $0 < \lambda < 1$

$$z(t) = \frac{1-\lambda^t}{1-\lambda} \underset{t\to\infty}{\rightarrow} \frac{1}{1-\lambda},$$

then the reachable set of the initial system

$$X_\varphi(t) = X_r(t) = \left\{x(t) : \|x\| \le \frac{1-\lambda^t}{1-\lambda}\right\}$$

corresponds to our intuitive considerations on this symmetric system.

4 Application to the Linear Control System in the Hilbert Space

Consider a particular case of system (1), assuming that it is linear stationary with respect to x and acts in a the Hilbert Space H:

$$x(t+1) = Lx(t) + b(u(t)), u \in U, \tag{7}$$

where L is a linear self-adjoint operator, $b : U_O \to H$, $U \subset U_O$, and $b(U)$ is a compact set in H.

We take the family of extending maps, $\varphi_\alpha = \langle \Psi_\alpha, x\rangle$, $\alpha \in A$ with Ψ_α, and A to be determined, and identify the corresponding family of extensions:

$$y_\alpha(t+1) = \langle \Psi_\alpha, Lx(t) + b(u(t))\rangle = \langle x(t), L\Psi_\alpha\rangle + \langle \Psi_\alpha, bu\rangle .$$

Then we assume that condition (4) is satisfied with

$$C_\alpha(z) = \lambda_\alpha z + \nu_\alpha \triangleq \lambda_\alpha z + \sup_{u\in U}\langle \Psi_\alpha, bu\rangle ,$$

where λ_α is some arbitrary real number.

To ensure this last condition it is sufficient to take Ψ_α such that $L\Psi_\alpha = \lambda_\alpha \Psi_\alpha, \alpha = 1, 2, \ldots$.

We suppose that all eigenvalues of L satisfy the condition $|\lambda_\alpha| < 1$. In this case for each λ_α we have a discrete scalar process

$$z(t+1) = \lambda_\alpha z(t) + \nu_\alpha$$

which converges to $z_\alpha^+ = (1-\lambda_\alpha)^{-1}\nu_\alpha$ when $t \to \infty$.

Repeat these steps for $\varphi_\alpha = -\langle \Psi_\alpha, x\rangle$ to receive $z_\alpha^- = (1+\lambda_\alpha)^{-1}\nu_\alpha^-$. After that we can note the final result:

$$X_\varphi = \bigcap\{x : z_\alpha^- \le \langle \Psi_\alpha, x\rangle \le z_\alpha^+\}$$

which is an external estimate of the invariant set of the system (7).

Example 2.[1] Consider the discrete case of the first initial boundary-value problem

$$x(t+1,\xi) = x(t,\xi) + h\frac{\partial^2 x(t,\xi)}{\partial \xi^2} + u(t,\xi) \quad t \in T = \{0,1,\ldots\}, 0 \le \xi \le \pi , \tag{8}$$

[1] Prepared by D. Rosenraukh, Irkutsk Computer Center, USSR.

$$x(t,0) = x(t,\pi) = 0, x(0,\xi) = \sin\xi, \|u\| \leq 1 \ . \tag{9}$$

Let the unbounded operator L be defined by the functions from $C^\infty(0,\pi)$ with the property (9) (set $D(L)$). Assume that it corresponds to the differential operator $\left(1 + h\frac{\partial^2}{\partial\xi^2}\right)$ on $[0,\pi]$, where $x(t,\xi)$ is determined for each $t \in T$ as an element from $L_2(0,\pi)$ and

$$\langle Lp, q\rangle = \int_O^\pi \left(p + h\frac{\partial^2 p}{\partial\xi^2}\right) q d\xi = \langle p, Lq\rangle, \forall p, q \in D(L) \ ,$$

i.e., L is a self-adjoint operator.

Since $\lambda_\alpha = 1 - \alpha^2 h$ and $\Psi_\alpha(\xi) = \sin\alpha\xi$, $\alpha = 1, 2, \ldots$, are the eigenvalue and eigenvector of the operator L, respectively, then

$$\phi_\alpha = \langle\Psi_\alpha, x\rangle = \int_O^\pi \Psi_\alpha(\eta) x(t,\eta) d\eta, y_\alpha(t) = \phi_\alpha(x(t,\xi))$$

and

$$X_\phi = \bigcap_{\alpha\in A} \{x(t,\xi) : z_\alpha^*(t) \leq \int_O^\pi \Psi_\alpha(\eta) x(t,\eta) d\eta \leq z_{\alpha *}(t)\}, z_\alpha^*$$

is determined by

$$z_\alpha^*(t+1) = (1 - \alpha^2 h) z_\alpha^*(t) + (\pi/2)^{\frac{1}{2}} \ ,$$

$$z_\alpha^*(0) = \langle\Psi_\alpha, \sin\xi\rangle$$

and $z_{\alpha *}$ can be found by using $\phi_\alpha = -\langle\Psi_\alpha, x\rangle$.

Acknowledgment

The main part of this work was developed by the author during his visit in April–June 1989 at the Mathematical Branch of the International School for Advanced Studies (ISAS), Trieste, Italy. The author wishes to thank Professor A. Cellina for the discussion of the paper and for the excellent working conditions at ISAS.

References

Krotov, V.F., Gurman, V.I. (1973). *Methods and Problems of Optimal Control*, Nauka, Moscow (in Russian).

Moskalenko, A.I. (1983). *Methods of Nonlinear Mappings in Optimal Control*, Nauka, Novosibirsk (in Russian).

Krotov, V. (1988). A technique of global bounds in optimal control theory, *Control and Cybernetics*, 17 (2–3).

Gurman, V.I. (1985). *Extension Principle in Control Problems*, Nauka, Moscow (in Russian).

Konstantinov, G.N. (1983). *Normalizing of the Dynamic Systems Inputs*, Irkutsk University, Irkutsk (in Russian).

Tolstonogov, A.A. (1986). *Differential Inclusions in Banach Space*, Nauka, Novosibirsk (in Russian).

CONTROLLING THE DYNAMICS OF SCALAR REACTION DIFFUSION EQUATIONS BY FINITE DIMENSIONAL CONTROLLERS

Pavol Brunovský
Institute of Applied Mathematics, Comenius University
Mlynská dolina, 84215 Bratislava, Czechoslovakia

The problem of stabilizing equilibria of steady states of distributed parameter systems by finite dimensional (i.e. having a finite number of inputs and outputs) controllers has been widely studied lately. In this paper we address a more general problem which contains the stabilization one as its special case. We are interested in finding a finite dimensional (in the above sense) feedback control which, if added to a given system, would make the essential dynamics of the resulting system to be equal to a one prescribed in advance.

In order to formulate our problem precisely we have to specify what we mean by "essential". To this end we first introduce the class of systems for which it is meaningful.

Consider an abstract ordinary differential equation in a Banach space X in the setting of [4]

$$\dot{y} + Ay = F(y) \tag{1}$$

where

(a) A is a sectorial (in general unbounded) operator [4].

(b) $F \in C^1(X^\alpha, X) \cap Lip_L(X^\alpha, X)$, $0 \leq \alpha < 1$.

(by $Lip_L(X^\alpha, X)$ we denote the space of function from X^α to X with Lipschitz constant L endowed by the C^0 topology; for the definition of the fractional space X^α cf.[4]).

Note that the abstract equation (1) includes reaction diffusion equations

$$u_t = \Delta u + f(x, u, \nabla u)$$

on bounded domains with sufficiently smooth boundaries and appropriate boundary conditions, f satisfying certain regularity and growth conditions, as well as certain systems of such equations.

In [4] it is proved that under the conditions (a), (b)

(i) $-A$ generates a strongly continuous semigroup e^{-At}

(ii) The equation (1) generates a C^1-semiflow S_t on X^α defined by $S_t(y_0) = y(t)$, where $y(t)$ is the solution of (1) satisfying $y(0) = y_0$.

(iii) The variation of constants formula

$$S_t(y) = e^{-tA}y + \int_0^t e^{-(t-s)A}F(S_s(y))ds$$

holds.

In general, the semiflow S in not invertible. However, it has been observed that in many cases there is a finite dimensional invariant manifold the restriction of S to which can be extended to a flow on $\mathbf{R}$ and the manifold attracts all the trajectories of S. The existence and the properties of such manifolds (called *inertial*) have been extensively studied lately (see e. g. [1,2,3]).

A situation in which such a manifold exists is the presence of a sufficiently large gap in the spectrum $\sigma(A)$ of A. Assume that $\sigma(A)$ admits a spectral decomposition at $\mu > 0$ with gap 2η, i.e.

$$\sigma(A) = \sigma_1 \cup \sigma_2$$

where

$$\sigma_1 = \{\lambda \in \sigma(A) : Re\lambda < \mu - \eta\}$$
$$\sigma_2 = \{\lambda \in \sigma(A) : Re\lambda > \mu + \eta\}.$$

Denote P_i the spectral projection corresponding to σ_i, $X_i := Range\ P_i$, $A_i := AP_i = P_iA$, $F_i := P_iF$, $y_i := P_iF$, $i := 1,2$. Then, A_1 is bounded and the following estimates hold:

$$| e^{-A_1 t} P_1 | \leq M e^{-(\mu-\eta)t} \ for\ t \leq 0,$$
$$| e^{-A_2 t} P_2 | \leq N e^{-(\mu+\eta)t} \ for\ t \geq 0,$$
$$| e^{-A_2 t} P_2 y |_\alpha \leq N_\alpha t^{-\alpha} e^{-(\mu+\eta)t} | y | \ for\ t > 0$$

. The equation (1) can be written as a system of equations

$$\dot{y}_1 + A_1 y_1 = F_1(y_1, y_2),$$
$$\dot{y}_2 + A_2 y_2 = F_2(y_1, y_2). \qquad (2)$$

In general, the constants M, N, N_α depend on the place $\sigma(A)$ is partitioned. Nevertthless, in some important cases (e.g. if A is self-adjoint), M, N are indepndent on μ.

The inertial manifold theorem of [2] asserts that if L is small compared to the gap η then S has an invariant manifold $\mathcal{M}$ which is a graph of a globally Lipschitz continuous function $h \in C^1(X_1, X_2^\alpha)$, $X_2^\alpha := X_2 \cap X^\alpha$. We recall that by "invariant" we mean that for each $y_0 \in \mathcal{M}$ there is a curve $y : \mathbf{R} \to \mathcal{M}$ such that $y(t+\tau) = S_\tau y(t)$ for each $t \in \mathbf{R}$ and each $\tau \geq 0$. Another important property of $\mathcal{M}$ is exponential tracking: every trajectory $y(t)$ of S has its "shadow" $y_\mathcal{M}$ in $\mathcal{M}$ which is a trajectory of $S |_\mathcal{M}$ satisfying

$$e^{\mu t} | y(t) - y_\mathcal{M}(t) | \to 0 \ for\ t \to \infty.$$

Now, it is quite on hand, why one can consider the dynamics on $\mathcal{M}$ as essential: it is invertible and by attracting all outside trajectories with a high exponential rate governs the entire dynamics.

In case X_1 is finite dimensional, so is $\mathcal{M}$.This is true if e.g. A has a compact resolvent which is the case for reaction diffusion equations (2) with Dirichlet or Neumann boundary conditions. It is our goal to control the dynamics on $\mathcal{M}$ in such a case.

The dynamics of the uncontrolled equation on $\mathcal{M}$ is the dynamics of the ordinary differential equation

$$\dot{y}_1 + A_1 y_1 = F_1(y_1, h(y_1)) \qquad (3).$$

We are interested in finding a feedback $U : X_1 \to X_1$ which, added to the system (1), would bring the differential equation on the inertial manifold to the form

$$\dot{y}_1 + A_1 y_1 = \Phi(y_1),$$

where $\Phi : X_1 \to X_1$ has been chosen in advance.

In general, adding a feedback U to F may destroy or alter the inertial manifold. Therefore, we have to formulate our problem as follows:

Given $\Phi : X_1 \to X_1$ find $U : X_1 \to X_1$ such that if $\mathcal{M}_U := graph\ h_U$ is the inertial manifold for the system

$$\dot{y} = Ay + F(y) + U(y_1)$$

then

$$F_1(y_1, h_U(y_1)) + U(y_1) = \Phi(y_1) \tag{4}$$

for all $y_1 \in X_1$.

To solve this problem let us first outline a method of construction of inertial manifolds [2]. By C_μ we denote the Banach space of continuous functions $\phi : (-\infty, 0] \to X$ satisfying $sup_{t\leq 0} e^{\mu t}|\phi(t)| < \infty$ endowed by the norm

$$\|\phi\|_\mu = sup_{t\leq 0} e^{\mu t}|\phi(t)|_\alpha.$$

We define $T : X_1 \times C_\mu \to C_\mu$ by

$$T(\xi, \phi)(t) = e^{-A_1 t}\xi + \int_0^t e^{-A_1(t-s)} F_1(\phi(s))ds + \int_{-\infty}^t e^{-A_2(t-s)} F_2(\phi(s))ds.$$

One has $y = T(\xi, \phi)$ if and only if $y(t)$ is a solution of the linear nonhomogeneous equation

$$\dot{y} = Ay + F(\phi(t))$$

from C_μ satisfying $P_1 y(0) = \xi$. It follows that $y(t)$ is a solution of the nonlinear equation (1) from C_μ satisfying $P_1 y(0) = \xi$ if and only if y is a fixed point of $T(\xi, .)$. If

$$\nu := L[\frac{M+N}{\eta} + \frac{2-\alpha}{1-\alpha} N_\alpha \eta^{\alpha-1}] < 1, \tag{5}$$

then $T(\xi, .)$ is a contraction uniform in ξ and, therefore, has a unique fixed point y_ξ^* which is a C^1 function of ξ with Lipschitz constant $\frac{M}{\nu-1}$ [2] . The map $h : X_1 \to X_2$ given by $h(\xi) := P_2 y_\xi^*(0)$, i.e.,

$$h(\xi) = \int_{-\infty}^0 e^{A_2 s} F_2(y_\xi^*(s))ds$$

defines the C^1 inertial manifold $\mathcal{M}$ by

$$\mathcal{M} := graph\ h.$$

Note that $\mathcal{M}$ is homeomorphic to X_1.

We now show that under certain assumptions relating the constants M, N, N_α , η and μ the problem (4) has a solution. We note that a solution U of (4) is a fixed point of the map Ψ given by

$$\Psi(U)(y_1) := \Phi(y_1) - F_1(y_1, h_U(y_1))$$

To prove that (4) has a solution we show that Ψ is a contraction in an appropriate Banach space. To this end we include U as a parameter into the map T. We define $T : X_1 \times Lip_{KL}(X_1, X_1) \times C_\mu \to C_\mu$ (K to be determined later) by

$$T(\xi, U, \phi)(t) = e^{-A_1 t}\xi + \int_0^t e^{A_1(t-s)}[F_1(\phi(s)) + U(P_1\phi(s))]ds$$

$$+ \int_{-\infty}^t e^{-A_2(t-s)} F_2(\phi(s))ds$$

Let $U \in Lip_{KL}(X_1, X_1)$. Assume that

$$\tilde{\nu} := \tilde{L}(\frac{M+N}{\eta} + \frac{2-\alpha}{1-\alpha}N_\alpha \eta^{\alpha-1}) < \frac{1}{2}, \tag{6}$$

with $\tilde{L} := L(1 + K|P_1|)$, the Lipschitz constant of $F + UP_1$. Then, by the uniform contraction theorem, the fixed point $y^*_{\xi,U}$ of $T(\xi, U, .)$ is a Lipschitz continuous function of ξ with constant $2M$. Since $(y_1, h_U(y_1)) = y^*_{\xi,U}(0)$, a Lipschitz constant of $\Psi(U)$ with respect to ξ is $L(1 + 2M)$. Thus, if (6) is satified with $K := 1 + 2M$, Ψ maps $Lip_{KL}(X_1, X_1)$ into itself. Further, a Lipschitz constant of $y^*_{\xi,U}(0)$ with respect to U is $\frac{2M}{\mu-\eta}$ from which it follows that Ψ is a contraction provided

$$\frac{2ML}{\mu - \eta} < 1 \tag{7}$$

Applying the contraction mapping theorem we obtain the following result:

Let $\Phi \in Lip_L(X_1, X_1)$. Assume that the inequalities (6), (7) are satisfied with $K := 1 + 2M|P_1|$. Then, there is a unique $U \in Lip_{KL}(X_1, X_1)$ such that

$$F_1(y_1, h_U(y_1)) + U(y_1) = \Phi(y_1) \text{ for all } y_1 \text{ in } X_1,$$

i. e. , if the feedback $U(y_1)$ is added to the system (1) then the reduction of the semiflow of the resulting system to the inertial manifold $\mathcal{M}_U$ is given by the equation

$$\dot{y}_1 + A_1 y_1 = \Phi(y_1).$$

We now apply this result to the scalar reaction diffusion equation

$$y_t = y_{xx} + f(y), \; 0 \le x \le 1, t \ge 0$$

with $f \in Lip_L(\mathbf{R}, \mathbf{R})$. For simplicity we consider this equation with Dirichlet boundary conditions

$$y(t, 0) = y(t, 1) = 0$$

but the result applies to other separated linear boundary conditions as well.

To put this equation into the abstract framework we take $X := L_2(0, 1)$ and define $Ay := -y''$ for $y \in H_2 \cap H_0^1$, $F(y)(x) := f(y(x))$ for $0 \le x \le 1$, $\alpha := 0$. Then, A is sectorial and $F \in Lip_L(X_1, X_1)$.

We have $\sigma(A) = \{\lambda_n : n := 1, 2, 3, ...\}$ with $\lambda_n = n^2\pi^2$. Choosing $\mu := \frac{1}{2}(\lambda_n + \lambda_{n+1})$ we have $X_1 = span\{\phi_1, ..., \phi_n\}$, $X_2 = span\{\phi_{n+1}, ...\}$ with $\phi_n(x) = sin\ n\pi x$ being the eigenvalue of λ_n; we can take $\eta := n$. We have

$$|e^{-A_1 t}| \leq e^{-\lambda_n t}\ (< e^{-(\mu-\eta)t})\ for\ t \leq 0,$$

$$|e^{-A_2 t}| \leq e^{-\lambda_{n+1} t}\ (< e^{-(\mu+\eta)t})\ for\ t \geq 0.$$

Since $M = N = N_\alpha = 1$ independently of n, for any L we can choose n so large that the estimates (6),(7) are met. Consequently, for fixed L, there exists an $n > 0$ such that for $X_1 := span\{\phi_1, ..., \phi_n\}$ our abstract result applies.

Its application does not immediately give a complete freedom of the choice of the dynamics - the dimension of the inertial manifold depends on the Lipschitz constant of Φ. Since the Lipschitz constant of A_1 is equal to λ_n and, hence, increases with n, this seems to limit our influence on the dynamics on the inertial manifold considerably.

Neverthless, employing the dynamics of the uncontrolled equation we can do better. Choosing $\Phi \in Lip_L(\mathbf{R^k}, \mathbf{R^k})$ for a fixed k, we can construct an inertial manifold of dimension k on which the dynamics is given by the equation

$$\dot{z} = \Phi(z). \tag{8}$$

Indeed, choose n such that the estimates (6),(7) hold with L replaced by $L + \lambda_k$. Denote $X_1 := span\{\phi_1, ...\phi_n\} = Z_1 \oplus Z_2$, where

$$Z_1 := span\{\phi_1, ..., \phi_k\},$$

$$Z_2 := span\{\phi_{k+1}, ..., \phi_n\},$$

$B_i := A_1|_{Z_i}, y_i = (z_1, z_2)$. Define $\tilde{\Phi} : X_1 \to X_1$, $\tilde{\Phi}(y_1) := (\tilde{\Phi}^1(y_1), \tilde{\Phi}^2(y_1))$ by

$$\tilde{\Phi}^1(y_1) := B_1 z_1 + \Phi(z_1),$$

$$\tilde{\Phi}^2(y_2) := 0.$$

Since the Lipschitz constant of $\tilde{\Phi}$ is $\lambda_k + L$, there exists a feedback $U : X_1 \to X_1$ such that the dynamics of the controlled system on the inertial manifold is given by the equation

$$\dot{y}_1 + A_1 y_1 = \tilde{\Phi}_1(y_1)$$

which is

$$\dot{z}_1 = \Phi(z_1),$$

$$\dot{z}_2 + B_2 z_2 = 0.$$

We have $B_2 = diag\{\lambda_{k+1}, ..., \lambda_n\}$. Therefore, z_2 decays exponentially with rate λ_{k+1}. This means that in the inertial manifold $\mathcal{M}_U$ we have constructed an inertial submanifold of dimension k on which the dynamics is given by (7).

REFERENCES

1. A. I. Babin, M. I. Vishik : Attractors of evolutionary equations (in Russian). Nauka, Moscow 1989

2. S. N. Chow, K.Lu: Invariant manifolds for flows in Banach spaces. J. Diff. Equations 74(1988),285-215

3. C. Foias, G. R. Sell, C.Témam: Inertial manifolds for nonlinear evolutionary equations. J. Diff. Equations 73(1988), 309-353

4. D. Henry: Geometric Theory of Semilinear Parabolic Equations. Lecture Notes in Math. 840, Springer, New York 1981

ESTIMATION OF CATALYST PELLET ACTIVITY DISTRIBUTION

Alena Brunovská
Department of Organic Technology, Slovak Institute of Technology, Radlinskeho 9, 812 37 Bratislava, Czechoslovakia

Noble metal catalysts, such as platinum, palladium, silver and others, are widely used in industry for hydrogenation and oxidation reactions. Mostly they are composed by inert porous support in which noble metal catalysts are dispersed. The reacting gas has to diffuse into the interior of the pellet where chemical reaction proceeds. If the rate of reaction is small compared with the rate of diffusion, the concentration of reacting gas at the pellet centre is little different from that on the surface. On the other hand, when the rate of the reaction is large compared to the rate of diffusion, the concentration of reactant is depleted by the reaction before it has a chance to diffuse within the pellet and the catalyst in the interior is not being used to any extent. The ratio of the actual reaction rate to its value when there is no diffusion limitation is called effectiveness factor.

The performance of the catalyst pellets can be significantly improved through the use of nonuniform noble metal distribution. For example, for positive order isothermal reactions, diffusional resistance reduces the reactant concentration and effectiveness factor the maximal value of which is obtained by concentration of the active catalyst near the external surface. However, in reactions of negative order kinetics (or Langmuir - Hinshelwood kinetics), diffusion resistance can enhance the effectiveness factor and the best location of the active catalyst is inside the pellet. Another reason for nonuniform activity distribution is to increase selectivity and resistance against deactivation.

The catalyst pellets with nonuniform active catalyst distribution are prepared by impregnation of support materials by solutions containing a precursor of the active ingredient and innactive species which is adsorbed on the support.

The important components of catalyst design are :

- estimation of optimal activity distribution for a given process
- estimation of real activity distribution of produced pellets

The first estimation method we need at the beginning, before pellet preparation. The following step is to develop a method of preparation for a catalyst with desired activity distribution. To control this impregnation procedure we need an estimation method of activity distribution on produced pellets.

The optimal catalyst pellet activity distribution for maximizing the effectiveness factor or global selectivity as well as the global yield for general reaction networks with arbitrary kinetics and finite external heat and mass transfer resistances is presented by Wu et al. [1]. The optimal activity distribution for reacting systems which undergo deactivation is analysed in the paper [2]. In both papers a general optimality criterion has been developed, which allows to conclude that the optimal activity distribution is of the Dirac - delta type.

The estimation method of catalyst pellet activity distribution has been discussed in the papers [3, 4, 5]. In the papers [3, 4] the proposed estimation method has been verified on simulated data (in the paper [3] for a positive order testing reaction, in the paper [4] for a zero order testing reaction). Experimental data have been treated in the paper [5].

The optimality criterion for optimal activity distribution as well as the gradient in the case of pellet activity distribution estimation method have been developed with the help of the adjoint equations. In both cases the results have been checked by an independent method (in the first case by testing few examples numerically, in the second case by experiments). In this paper we summarize our main results.

Optimal catalyst pellet activity distribution for deactivating systems

The Optimization Problem

The catalyst which is progressively poisoned with operating time has to be periodically replaced or regenerated, depending upon whether the poisoning is irreversible or reversible. The duration of the operating time and the values of the effectiveness factor as a function of time depend upon the active catalyst distribution within the support. In general, by locating the active catalyst inside the pellet it is possible to increase the duration of the operating time. On the other hand, at least for positive order reactions, the maximum value of the effectiveness factor is obtained when the active catalyst is located at the external surface. This is why we use an economic criterion

$$\text{profit/time} = \frac{\text{price of the product - cost of the catalyst}}{\text{operating time}} = \frac{\alpha_1 \int_0^{\tau^*} \eta \, d\tau - \alpha_2}{\tau^*} \qquad (1)$$

where α_1 and α_2 are weighting coefficients proportional to the price of the product and to the cost of the catalyst, respectively, τ^* is the operating time and η is the effectiveness factor.

The aim of this work is to determine the initial pellet activity distribution $a(\varphi,0)$ and the operating time τ^* for which the maximum value of the following objective function, proportional to the profit per time, defined above

$$\mathcal{J}\,[a(\varphi,o),\tau^*] = \frac{\gamma \int_0^{\tau^*} \eta \, d\tau - 1}{\tau^*} \qquad (2)$$

($\gamma = \alpha_1/\alpha_2$) is obtained. We optimize over the class of all possible distributions of the same amount of active catalyst.

The Basic Equations

Let us consider a catalyst pellet in which an irreversible reaction is taking place together with irreversible adsorption of catalyst poison. Since the rate of the poison adsorption is usually considerably lower than that of the catalytic reaction (the form of which may otherwise be arbitrary) the quasi-steady-state approximation can be safely adopted. In addition, we assume negligible external resistances to mass and heat transport. The catalyst activity distribution is a function of location and time and is defined as the ratio between the local concentration of available catalytically active sites and its volume averaged initial value

$$a(\varphi,\tau) = \sigma(\varphi,\tau)/\bar{\sigma} \tag{3}$$

where
$$\bar{\sigma} = (n + 1) \int_0^1 \sigma(\varphi,0)\, \varphi^n\, d\varphi \tag{4}$$

Under these conditions, the model equations in dimensionless form are as follows:

Mass balance of the reactant

$$\nabla^2 Y = \Phi^2 R \tag{5}$$

Mass balance of the poison

$$\nabla^2 Y_p = \Phi_p^2 R_p \tag{6}$$

Energy balance

$$\nabla^2 \upsilon = -\beta \Phi^2 R \tag{7}$$

with boundary conditions

$$\varphi = 0 : \partial Y/\partial\varphi = \partial Y_p/\partial\varphi = \partial\upsilon/\partial\varphi = 0$$

$$\varphi = 1 : Y = Y_p = \upsilon = 1 \tag{8}$$

The deactivation reaction is accounted for by a balance of the active sites, which in terms of the activity distribution function reduces to

$$\frac{\partial a}{\partial \tau} = - R_p \tag{9}$$

with initial condition

$$a = a(\varphi,0) \text{ at } \tau = 0 \tag{10}$$

where the initial activity distribution has to satisfy the constraint which arise from its definition (3) and Eq.(4)

$$(n + 1) \int_0^1 a(\varphi,0)\, \varphi^n\, d\varphi = 1 \tag{11}$$

The rates of the reaction and the poisoning processes have the following general form

$$R = R(Y,Y_p,a,\upsilon) \; ; \; R_p = R_p(Y,Y_p,a,\upsilon) \tag{12}$$

The effectiveness factor η is normalized with respect to the initial value of the reaction rate computed at surface conditions and to the initial activity distribution

$$\eta = \int_0^1 \varphi^n R \, d\varphi / \int_0^1 a(\varphi,0)\, \varphi^n d\varphi = (n + 1) \int_0^1 \varphi^n R \, d\varphi = \bar{R} \qquad (13)$$

and is equal to the mean reaction rate.

General Condition for Optimal Activity Distribution

Consider the general deactivation process described above [eqs (5) - (7) and (9)] in a symmetric domain with boundary conditions (8) and initial condition (10). The goal is to find the initial distribution $a(\varphi,0)$ subject to the constraints

$$(n + 1) \int_0^1 \varphi^n \, a(\varphi,0) \, d\varphi = 1 \text{ and } a(\varphi,0) \geq 0 \qquad (14)$$

and the time $\hat{\tau} > 0$, such that for $\tau^* = \hat{\tau}$ and $a(\varphi,0) = a(\varphi,0)$ the objective function (2) is maximized .

In the paper [2] the following optimality criterion has been developed :*If* $a(\varphi,0)$, *is optimal, then, for any given initial distribution* $a(\varphi,0)$, *one has*

$$\int_0^1 \varphi^n \, \Psi(\varphi,0) \, a(\varphi,0) \, d\varphi \geq \int_0^1 \varphi^n \Psi(\varphi,0) \, a(\varphi,0) \, d\varphi \qquad (15)$$

where $\Psi(\varphi,\tau)$ *is obtained as a solution of the system of adjoint equations*

$$\nabla^2 p + \frac{\partial R}{\partial Y} (1 - p\Phi^2 + s\beta\Phi^2) - \frac{\partial R_P}{\partial Y}(q\Phi_p^2 + \Psi) = 0 \qquad (16)$$

$$\nabla^2 q + \frac{\partial R}{\partial Y_p} (1 - p\Phi^2 + s\beta\Phi^2) - \frac{\partial R_P}{\partial Y_p}(q\Phi_p^2 + \Psi) = 0 \qquad (17)$$

$$\nabla^2 s + \frac{\partial R}{\partial \upsilon} (1 - p\Phi^2 + s\beta\Phi^2) - \frac{\partial R_P}{\partial \upsilon}(q\Phi_p^2 + \Psi) = 0 \qquad (18)$$

$$\frac{\partial \Psi}{\partial \tau} + \frac{\partial R}{\partial a} (1 - p\Phi^2 + s\beta\Phi^2) - \frac{\partial R_P}{\partial a}(q\Phi_p^2 + \Psi) = 0 \qquad (19)$$

with the boundary and terminal conditions

$$\tau = \hat{\tau} \quad : \quad \Psi(\varphi,0) = 0 \tag{20}$$

$$\tau \in \langle 0,\hat{\tau}\rangle : \quad \varphi = 0 : \partial p/\partial\varphi = \partial q/\partial\varphi = \partial s/\partial\varphi = \partial\Psi/\partial\varphi = 0 \tag{21}$$

$$\varphi = 1 : p = q = s = 0 \tag{22}$$

with coefficients depending on $a(\varphi,0)$.

The optimality criterion (15) practically excludes any initial distribution $a(\varphi,0)$ which is not of the Dirac-delta type. In fact for any given distribution it is possible to construct a suitable Dirac-delta distribution which improves the objective functional (2). The maximal value of the integral criterion (15) is obtained when all active catalyst is concentrated at the point where the function $\Psi(\varphi,0)$ is maximal. In addition the criterion (15) may exclude some delta distributions as well and indicate in which direction to move the activity location point to find the optimal one.

Example

For illustration let us consider the case of an isothermal first order reaction with dimensionless rate equation

$$R = a\ Y \tag{23}$$

which occurs together with independent chemisorption of catalyst poison, leading to the following rate expression for the deactivation process :

$$R_p = a\ Y_p \tag{24}$$

We will consider two type of activity distributions :

~ Dirac-delta activity distribution located at the point φ_1, i.e.

$$a(\varphi,\tau) = a(\varphi,0)\ \mu(\tau) = \frac{\delta(\varphi - \varphi_1)}{(n+1)\ \varphi_1^n}\ \mu(\tau) \tag{25}$$

~ Step function activity distribution (i.e. uniformly active region between the points φ_1 and φ_2)

$$\begin{aligned} &\varphi \in \langle 0,\ \varphi_1) \text{ and } \varphi \in (\varphi_2,\ 1\rangle &&: a(\varphi,\tau) = 0 \\ &\varphi \in \langle \varphi_1,\varphi_2\rangle &&: a(\varphi,\tau) = \mu(\tau)/(\varphi_2^{n+1} - \varphi_1^{n+1}) \end{aligned} \tag{26}$$

where $\mu(0) = 1$. The system of the adjoint equations is

$$\nabla^2 p + a(1 - p\,\Phi^2) = 0 \tag{27}$$

$$\nabla^2 q - q\,\Phi_p^2 a - \Psi\, a = 0 \tag{28}$$

$$\partial\Psi/\partial\tau + Y(1 - p\,\Phi^2) - q\,\Phi_p^2\, Y_p - \Psi\, Y_p = 0 \tag{29}$$

with boundary and terminal conditions (20) - (22).

In the case of a Dirac-delta activity distribution the solution of the model equations as well as the expression of the objective function can be obtained in a closed form [2]. The objective function (2) becomes a function of two parameters : the active point location φ_1 and the operating time τ^*. The model equations for a step function activity distribution have to be solved numerically. The objective function is a function of three parameters : φ_1, φ_2 and τ^*. The adjoint variable profiles $\Psi(\varphi,0)$ have been obtained by solving numerically the system of adjoint equations. One example for parameter values : $\alpha = 10$, $\gamma = 5$, $\Phi^2 = 1$, $n = 1$ is exhibited in Figs. 1 and 2 .

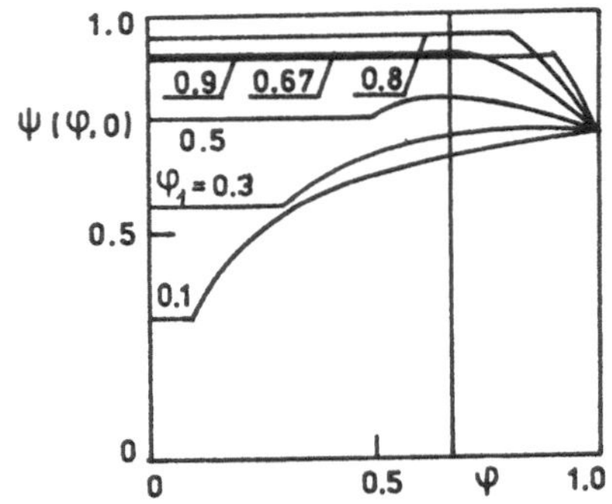

Fig.1 Dirac-delta activity distributions

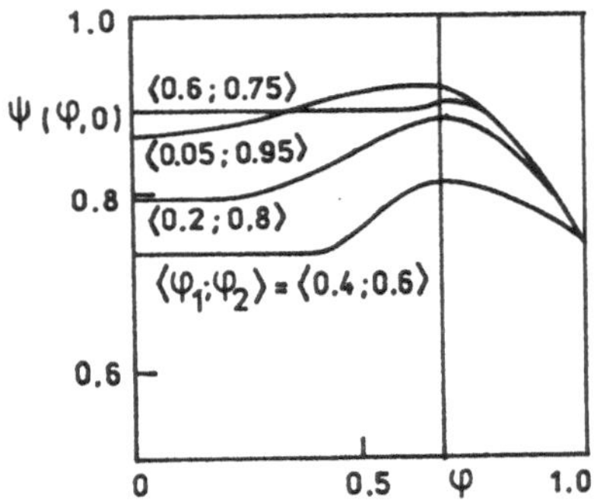

Fig.2 Step function activity distributions

The optimal Dirac-delta distribution, obtained by a standard optimization method, is located at $\varphi_1 = 0.67$ (as indicated by the solid vertical line). It is rather surprising that for all considered step-size distributions, even the widest one <0.05, 0.95>, the maximum of the $\Psi(\varphi,0)$ curves is very close to the location of the optimal Dirac-delta distribution. This provides a useful initial information for the optimum search. In Fig. 1 the adjoint profiles $\Psi(\varphi,0)$ are shown relative to Dirac delta distributions centered at various locations φ_1. It appears that using the criterion (15) it is possible to exclude the locations φ_1= 0.1, 0.3 and 0.5, since the corresponding adjoint functions exhibit their maximum values at other locations. In addition, the function $\Psi(\varphi,0)$ indicates in all cases, that the optimal location should be to the right (i.e. larger values φ_1) since the value of the integral in the right hand side of condition (15) increases when moving the Dirac delta location in this direction. On the other hand, the criterion (15) is not fine enough to exclude the location points to the right of the optimal one (i.e., 0.8 and 0.9). The only way to exclude such points is in fact by comparing the corresponding values of the objective function.

Estimation of the catalyst pellet activity distribution from kinetic data

The estimation of the activity distribution inside the catalyst pellet is useful for several reasons. One reason is to control the impregnation procedure, another one is to obtain some information about the mechanism of deactivation.

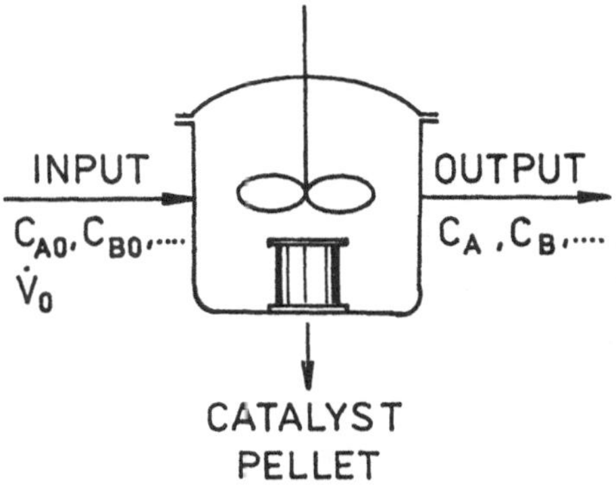

Fig. 3 Kinetic measurenments

The goal of the presented method is to estimate the activity distribution from kinetic data. The kinetic data have been obtained by measuring the stirred tank reactor outlet concentration of a fixed testing reaction for various feed rates (Fig. 3). The form of the testing reaction rate expression has been assumed to be known and the value of the reaction rate constants can be evaluated from measurements in the kinetic region on the crushed pellet. Also the values of the diffusion coefficients have to be estimated by a different method. For most of the testing reactions, the estimation of its value along with the activity distribution appears to be an ill-posed problem. In any case it is helpful to have some more information about the activity, e.g. which part of the pellet is active, or whether the distribution has increasing or decreasing tendency, etc.

Further assumption is that the system response is sensitive enough to the activity distribution. It depends on the testing reaction choice and on the experimental conditions. This can be tested by computer simulation of the pellet behaviour.

The proposed method has been tested on experimental data. The investigated pellet (γ-alumina impregnated with Pt) had a narrow region activity distribution. As the testing reaction hydrogenation of ethylene (Langmuir - Hinshelwood reaction rate expression) has been chosed. The estimated activity profile has been compared with the distribution of Pt obtained by the scanning electron microscope combined with energy - dispersive analyser of X-rays. The experiment and catalyst pellet preparation is described in detail in the paper [5].

Optimization Problem

Let us consider a catalyst pellet in a continuous stirred reactor in which the testing reaction

$$A + B \Rightarrow products$$

takes place. The dependence of the reactor outlet composition on feed rate under steady state conditions is measured.

The problem is to estimate the pellet activity distribution to obtain the best fit of the measured and the computed reactor outlet concentrations. As the objective function

$$F[a(\varphi)] = \sum_{i=1}^{I} (Y_{Ai} - Y_{Ai,exp})^2 = \min \tag{30}$$

has been used, where Y_{Ai} is the dimensionless reactor outlet concentration corresponding to the activity distribution $a(\varphi)$, $Y_{Ai,exp}$ is the experimental dimensionless concentration and I is the number of measurements.

The activity distribution has been considered as a piecewise linear function given by the values $a(\varphi_k)$ in equidistant mesh points dividing the interval $\langle 0,1\rangle$.

The Basic Equation

Let us assume perfect gas-to-solid mass transfer, constant temperature in the active layer and constant diffusion coefficients. Then the model dimensionless equations are as follows :

Pellet mass balances

$$\nabla^2 Y_A = \Phi^2 R \tag{31}$$

$$\nabla^2 Y_B = \delta\, \Phi^2 R \tag{32}$$

boundary conditions

$$\varphi = 0 \quad : \quad dY_A/d\varphi = dY_B/d\varphi = 0 \tag{33}$$

$$\varphi = 1 \quad : \quad Y_A = Y_A(1)$$

$$Y_B = Y_B(1) \tag{34}$$

Reactor mass balances

$$1 - Y_A(1) = Z_{RA}\bar{R} \tag{35}$$

$$1 - Y_B(1) = Z_{RB}\bar{R} \tag{36}$$

Reaction rate equation

$$R = a^2 \frac{\omega^2 Y_A Y_B}{(1 + \varkappa_1 Y_A + \varkappa_2 Y_B)^2} = a^2 \xi(Y_A, Y_B) \tag{37}$$

Mean reaction rate expression

$$\bar{R} = (n + 1) \int_0^1 R\, \varphi^n d\varphi \tag{38}$$

From Eqs (31) - (36) and (38) we obtain the boundary conditions

$$\varphi = 1 : \quad dY_A/d\varphi = \frac{\Phi^2}{Z_{RA}\,(n + 1)} [1 - Y_A(1)] \tag{39}$$

$$dY_B/d\varphi = \frac{\Phi^2\ \delta}{Z_{RB}(n+1)}\ [1 - Y_B(1)] \tag{40}$$

and the following relation between concentrations of components *A* and *B*

$$Y_B = Y_B(1) - \delta\ [Y_A(1) - Y_A] \tag{41}$$

Using Eq. (41) we can reduce the system of model equations by considering the equations for the component *A* only. The system of model equations has been solved numerically.

Optimization Technique

The problem is to minimize the objective function (30) under the constraints

$$(n + 1) \int_0^1 a(\varphi)\ \varphi^n\ d\varphi = 1 \tag{42}$$

and non-negativity of a. The objective function has K parameters $a(\varphi_1), a(\varphi_2), \ldots\ldots, a(\varphi_K)$. To find the minimum of the function (30) a gradient type method has been employed. The gradient is the vector with the components

$$\frac{\partial F}{\partial a(\varphi_k)} = \sum_{i=1}^{I} \{-2(n+1)\ \Phi^2 \int_{\varphi_i-\Delta\varphi/2}^{\varphi_i+\Delta\varphi/2} \varphi^n (p_{Ai} + \delta p_{Bi})\ \xi_i d\varphi\ \}$$

$$k = 1, 2, \ldots, K \tag{43}$$

and p_A, p_B are the adjoint variables, which solve the adjoint equations

$$\nabla^2 p_A = (p_A + \delta p_B)\Phi^2 a^2\ \frac{\partial \xi}{\partial Y_A} \tag{44}$$

$$\nabla^2 p_B = (p_A + \delta p_B)\Phi^2 a^2\ \frac{\partial \xi}{\partial Y_B} \tag{45}$$

with boundary conditions

$$\varphi = 0\ :\ dp_A/d\varphi = dp_B/d\varphi = 0 \tag{46}$$

$$\varphi = 1\ :\ dp_A/d\varphi = [Y_A(1) - Y_{A,exp} - p_A(1)\Phi^2/Z_{RA}]/(n+1) \tag{47}$$

$$dp_B/d\varphi = [-p_B(1)\ \delta\Phi^2/Z_{RB}]/(n+1) \tag{48}$$

The derivation of the gradient components and the system of the adjoint equations is similar as for the n-th order reaction (see Appendix of the paper [3]).

Because of the constraint (42) and the non-negativity of $a(\varphi_k)$ the projected gradient method has been used. The (m+1)-th iteration has been computed from the m-th one by the scheme

$$a(\varphi_k) = \frac{a(\varphi_k)^m - \lambda^m y_k}{(n+1)\int_0^1 (a(\varphi_k)^m - \lambda^m y_k)\varphi^n d\varphi} \qquad (49)$$

where

$$y_k = \begin{cases} \partial F/\partial a(\varphi_k) & \text{for } a(\varphi_k)>\varepsilon \text{ or for } a(\varphi_k)\le \varepsilon \text{ and } \partial F/\partial a(\varphi_k)\le 0 \\ 0 & \text{for } a(\varphi_k)\le\varepsilon \text{ and } \partial F/\partial a(\varphi_k)>0 \end{cases} \qquad (50)$$

and the step length λ^m has been determined by a one-parameter optimization procedure in the gradient direction (the method of steepest descent). As the first approximation of the activity distribution the parabolic function has been chosen.

Results

The estimated activity distribution is in Fig. 4. In this figure the resulting activity distribution is compared with the normalized Pt distribution (ratio of Pt amount and total Pt amount in the pellet) from the scanning electron microscope. We note, that this two profiles

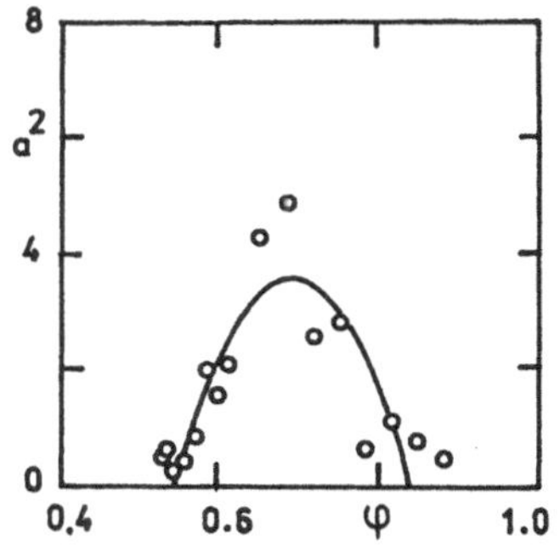

Fig. 4 Comparison of estimated activity distribution (————) and Pt distribution (◦).

we can compare just qualitatively, because there is not simple and known relation between activity distribution and active catalyst distribution which depends on several physical properties. The deacrease of the objective function during the iterative procedure indicate that the system is sensitive enough on the pellet activity distribution and that the choice of the testing reaction and experimental conditions appers to be suitable. The proposed gradient type estimation method works well for simulated [3, 4] as well as for experimental data.

List of symbols

a activity
a characteristic dimension of catalyst pellet
a_P equilibrium poison adsorbed amount
C concentration
D diffusion coefficient
F objective function in problem 2
$(-\Delta H)$ heat of reaction
$\mathcal{J}$ objective function in problem 1
n integer characteristic of pellet geometry (n=0, slab; n=1, cylinder; n=2, sphere)
p adjoint variable
q adjoint variable
r reaction rate
R dimensionless main reaction rate
R_P dimensionless poisoning rate
s adjoint variable
t time
T temperature
t^0 characteristic deactivation time
$\dot{V}$ volumteric flow rate
W catalyst mass
Y $= C/C_0$, dimensionless concentration
Z_{RA} $= r_0 W/(\dot{V} C_{A0})$, dimensionless parameter
Z_{RB} $= r_0 W/(\dot{V} C_{B0})$, dimensionless parameter

Greek letters

α $= \Phi_p^{\,2}/\Phi^2$, ratio of Thiele moduli

α_1 price of product

α_2 cost of catalyst

β $= (-\Delta H) D_A C_{A0}/(\Lambda T_0)$, dimensionless reaction heat

γ $= \alpha_1/\alpha_2$, dimensionless parameter

δ $= D_{eA} C_{A0}/(D_{eB} C_{B0})$, dimensionless parameter

η effectiveness factor

$\varkappa_1$ $= K_A C_{A0}$, dimensionless parameter

$\varkappa_2$ $= K_B C_{B0}$, dimensionless parameter

ξ concentration term in dimensionless reaction rate

ε accuracy

μ relative activity

τ $= t/t^0$, dimensionless time

Λ thermal conductuvity

λ step length in gradient method

φ dimensionless space coordinate

Φ $= a[r_0/(D_A C_{A0})]^{1/2}$, reaction Thiele modulus

Φ_p $= a[a_p/(D_p C_{p0} t^0)]^{1/2}$, poison Thiele modulus

Ψ adjoint variable

ω $= 1 + \varkappa_1 + \varkappa_2$ rate equation parameter

υ $= T/T_0$, dimensionless temperature

σ concentration of available catalytically active sites

Subscripts

* terminal conditions

0 feed stream conditions

A,B reactants

exp experimental points

p poison

1,2 activity location

References

1. Wu Hua, Brunovská A., Morbidelli M., Varma A., Optimal catalyst activity profiles in pellets : Nonisothermal general reacting systems with arbitrary kinetics, Chem. Engng. Sci., in press
2. Brunovská A., Morbidelli M., Brunovský P., Optimal catalyst pellet activity distribution for deactivating systems, Chem. Engng. Sci., in press
3. Brunovská A., Horák J., Estimation of the catalyst pellet activity distribution from kinetic data, Collection Czechoslovak Chem. Commun. 52, 2412 (1987)
4. Brunovská A., Horák J., Estimation of the catalyst pellet activity distribution from zero order kinetic data, Collection Czechoslovak Chem. Commun. 52, 2426 (1987)
5. Brunovská A., Remiarová B., Lebrun C., Estimation of the catalyst pellet activity distribution from experimental kinetic data, Collection Czechoslovak Chem. Commun. 54, 388 (1989)

Some remarks on the periodic linear quadratic regulator problem.

G.Da Prato [1]
Scuola Normale Superiore
56126 Pisa, Italy

1 Introduction and notation.

If X is a Hilbert space ,we shall denote by $L^2_\#(X)$, the set of all 2π-periodic mappings u: $\mathbf{R}\to X$, locally square integrable, and by $C_\#(X)$, the space of all 2π-periodic continuous mappings from $\mathbf{R}$ into X.

We consider here three Hilbert spaces, H(the *state* space), U(the *control* space) and Y(the *observations* space), and a dynamical system governed by the state equation :

$$\begin{cases} y'(t) = A(t)y(t)+B(t)u(t)+f(t) \\ \\ y(0) = y(2\pi) \end{cases} \tag{1.1}$$

We assume :

(H1) *For all* $t\in\mathbf{R}$, B(t) *is a linear bounded operator from* U *into* H. *Moreover* $B(t+2\pi) = B(t)$ *and* $B(\cdot)u$ *is continuous for any* $u\in U$.

(H2) *For all* $t\in\mathbf{R}$, $A(t)$: $D(A(t))\subset H\to H$ *generates a strongly continuous semigroup in* H. *Moreover* , $A(t+2\pi) = A(t)$ *and there exists a strongly continuous mapping* $G_A(\cdot,\cdot)$: $\{(t,s)\in\mathbf{R}^2\text{: } t\geq s\}\to \mathcal{L}(H)$, *such that*

$$\frac{\partial}{\partial t}G_A(t,s)x = A(t)G_A(t,s)x \text{ and } G_A(s,s)x = x \text{ for all } x\in H \text{ and } t>s.$$

(H3) *We have* $\lim_{n\to\infty} G_{A_n}(t,s)x = G_A(t,s)x$ *for all* $x\in H$, *uniformly on the bounded sets of* $\{(t,s)\in\mathbf{R}^2\text{: } t\geq s\}$,*where* $G_{A_n}(t,s)$ *is the evolution o operator generated by the Yosida approximations* $nA(t)(n-A(t))^{-1}$ *of* A(t).

---

[1] Work partially supported by the Italian National Project M.P.I. 40% " Equazioni di Evoluzione e Applicazioni Fisico-Matematiche"

(H4) $\qquad f \in L^2_\#(H)$.

Hypotheses (H2) and (H3) are fulfilled in many problems both parabolic and hyperbolic (see for instance [1], [7], [8]).

In (1.1) y(t) represents the *state* and u(t) the *control* of our system. A *mild solution* of (1.1) is a mapping $y \in C_\#(H)$ such that , for all interval $[a,b] \subset \mathbf{R}$, one has

$$(1.2) \qquad y(t) = G_A(t,a)y(a) + \int_a^t G_A(t,s)[B(s)u(s)+f(s)]ds \ ; \quad t \in [a,b]$$

If 1 belongs to the resolvent set $\rho(G_A(2\pi,0))$ of $G_A(2\pi,0)$, we say that A(t) is *non resonant* ; in this case, as well known, for any $u \in L^2_\#(U)$ there exists a unique mild solution of (1.1) given by :

$$(1.3) \qquad y(t) = G_A(t,0)(1-G_A(2\pi,0))^{-1} \int_0^{2\pi} G_A(2\pi,s)[B(s)u(s)+f(s)]ds$$

$$+ \int_0^t G_A(t,s)[B(s)u(s)+f(s)]ds$$

We recall that the non zero eigenvalues of $G_A(2\pi,0)$ are called the *Floquet exponents* of A.

We are also interested in the case when A(t) is resonant; in this case, for any $u \in L^2_\#(U)$ we set :

$$(1.4) \qquad \Lambda_u = \{y \in C_\#(H) \text{: y is a mild solution of (1.1)}\}$$

If Λ_u is non empty, the control u is said to be *admissible. We shall denote by* U_{ad} the set of all admissible controls

$$(1.5) \qquad U_{ad} = \{u \in L^2_\#(U) \ ; \ \Lambda_u \neq \varnothing\}$$

Our goal is to minimize the cost functional :

$$(1.6) \qquad J(u,y) = \int_0^{2\pi} \{\|C(t)y(t)\|^2 + \|u(t)\|^2\} dt$$

defined for all $u \in U_{ad}$ and $y \in \Lambda_u$. We assume :

(H5) *For all* $t \in \mathbf{R}$, $C(t)$ *is a linear bounded operator from* Y *into* H. *Moreover* $C(t+2\pi) = C(t)$ *and* $C(\cdot)x$ *is continuous for any* $x \in H$.

If there exist $u^* \in U_{ad}$ and $y^* \in \Lambda_{u^*}$ such that :

$$J(u^*,y^*) \leq J(u,y), \quad \text{for all } u \in U_{ad} \text{ and } y \in \Lambda_u$$

we say that u^* is an *optimal control* and y^* an *optimal state*.

In order to show the existence of an optimal control we will proceed as follows :

Step 1. We assume that (A,B) is *stabilizable* with respect to the observation C, that is :

(H6) *For all* $x \in H$ *there exists* $u \in L^2(0,\infty;U)$ *such that* $\int_0^{+\infty} \{\|C(t)y(t)\|^2 + \|u(t)\|^2\} dt < \infty$

In [6] was proved that this condition is necessary and sufficient for the existence of a positive 2π-periodic solution $Q_\#(t)$ of the Riccati equation

$$Q' + A^*Q + QA - QBB^*Q + C^*C = 0 \tag{1.7}$$

Step 2. We solve , under suitable assumptions, the dual problem :

$$\begin{cases} r'(t) + F^*(t)r(t) + Q_\#(t)f(t) = 0 \\ r(0) = r(2\pi) \end{cases} \tag{1.8}$$

and the closed loop equation

$$\begin{cases} y'(t) = F(t)y(t) - B(t)B^*(t)r(t) + f(t) \\ y(0) = y(2\pi) \end{cases} \tag{1.9}$$

where $F(t) = A(t) - B(t)B^*(t)Q_\#(t)$ is the *feedback operator.*

In order to solve problems (1.8) and (1.9) one requires that F(t) is non resonant. As shown in [5], this happens if (A,C) is detectable ([5]); a more general condition will be discussed in Section 3.

Step 3. By proceeding as in [5] and [4] , one first proves the identity :

(1.10) $$J(u,y) = J^* + \int_0^{2\pi} \|B^*Qy+B^*r+u\|^2 dt \; ; \text{ for all } u\in U_{ad} \text{ and } y\in \Lambda_u$$

where

(1.11) $$J^* = \int_0^{2\pi} \{<r,f>-\|B^*r\|^2\} dt$$

and r is the solution of (1.8).

By using (1.10) it is not difficult to prove that there exists an optimal control u* and that the optimal cost is given by J*.

A key point of the above program is to show that F(t) is non resonant. A result in this direction was given in [4]. In Section 2 we recall some results on Riccati equation and in Section 3 we will present some new result on the relation bertween the Floquet exponents of A and F.

2 Riccati equation.

We first introduce some notation. We set

$$\Sigma(H) = \{S\in \mathcal{L}(H) \; ; \; S \text{ is hermitian } \}; \;\; \Sigma^+(H) = \{S\in \Sigma(H) \; ; \; <Sx,x>\geq 0 \;\; \forall x\in H \}$$

We denote by $C_s([0,T];\Sigma(H))$ (resp.$C_s^{\#}(\Sigma(H))$) the set of all strongly continuous mappings $F:[0,T]\to\Sigma(H)$ (resp. the set of all strongly continuous mappings $F:\mathbf{R}\to\Sigma(H)$ which are periodic)

Next we assume (H1)-(H6), fix T>0 and consider the Cauchy problem :

(2.1) $$\begin{cases} \dfrac{dQ}{dt}+A^*Q+QA-QBB^*Q+C^*C = 0 \\ Q(T) = P_0 \end{cases}$$

where $P_0\in \Sigma^+(H)$, and the approximating problem :

(2.2) $$\begin{cases} \dfrac{dQ_n}{dt}+A_n^*Q_n+Q_nA_n-Q_nBB^*Q_n+C^*C = 0 \\ Q_n(T) = P_0 \end{cases}$$

where $A_n = nA(n-A)^{-1}$ is the Yosida approximation of A

We say that $Q\in C_s([0,T];\Sigma(H))$ is a *mild solution* of problem (2.1) if

$$(2.3)\qquad Q(t)x = e^{(T-t)A^*}P_0e^{(T-t)A}x + \int_t^T e^{(s-t)A^*}\{C^*C-P(s)BB^*P(s)\}e^{(s-t)A}x\,ds,$$

for all $x\in H, t\in[0,T]$.

The following result can be proved as in [2]

Proposition 2.1. *Assume* (H1), (H2), (H3) *and* (H5). *Then problem* (2.1) *has a unique mild solution* $Q\in C_s([0,T];\Sigma(H))$, *such that* $Q(t)\geq 0$ *for all* $t\in[0,T]$. *Moreover, problem* (2.2) *has a unique solution* Q_n *and, for all* $x\in H$

$$(2.4)\qquad \lim_{n\to\infty} Q_n(t)x = Q(t)x, \text{ uniformly in } t \text{ in } [0,T].$$

We consider now Riccati equation

$$(2.5)\qquad Q'+A^*Q+QA-QBB^*Q+C^*C=0$$

We say that $Q\in C_s^{\#}(\Sigma(H))$ is a periodic solution of (2.5) if, for any interval $[a,b]\subset \mathbf{R}$ one has

$$(2.6)\qquad Q(t)x = e^{(b-t)A^*}Q(a)e^{(b-t)A}x + \int_t^b e^{(s-t)A^*}\{C^*C-P(s)BB^*P(s)\}e^{(s-t)A}x\,ds,$$

for all $t\in[a,b]$ and $x\in H$.

The following result can be proved as in [6]

Proposition 2.2. *Assume* (H1), (H2), (H3),(H5) *and* (H6). *Let* $Q^{(\tau)}$ *be the mild solution to the problem*

$$(2.7)\qquad \begin{cases} \dfrac{dQ^{(\tau)}}{dt} + A^*Q^{(\tau)}+Q^{(\tau)}A-Q^{(\tau)}BB^*Q^{(\tau)}+C^*C=0 \\ Q^{(\tau)}(\tau)=0\ , t\leq\tau \end{cases}$$

Then for any $t\in\mathbf{R}$, $Q^{(\tau)}(t)$ *is increasing in* τ *and there exists the strong limit*

$$(2.8)\qquad \lim_{\tau\to\infty} Q^{(\tau)}(t)x := Q_{\#}(t)x \quad ; \ \forall x\in H$$

Moreover $Q_{\#}$ *is a periodic solution of* (2.5).

The operator $Q_\#$, defined by (2.8) is the *minimal nonnegative periodic solution* of Riccati equation (2.5).

3 Spectral properties of the feedback operator.

We assume here (H1)-(H6). We denote by $Q_\#$ the minimal nonnegative periodic solution of Riccati equation (2.5). We set

(3.1) $$F_n(t) = A_n(t) - B(t)B^*(t)Q_\#(t), \quad t\in \mathbf{R}.$$

where $A_n(t)$ is the Yosida approximation of $A(t)$, and denote by $G_{F_n}(t,s)$ the evolution operators generated by $F_n(t)$. Recalling hypothesis (H3), it is not difficult to show that

(3.2) $$\lim_{n\to\infty} G_{A_n}(t,s)x = G_A(t,s)x, \quad \text{for all } x\in H,$$

uniformly on the bounded sets of $\{(t,s)\in \mathbf{R}^2: t\geq s\}$.

The main result of this paper is the following :

Theorem 3.1. *Assume* (H1)-(H6); *then the following statements are equivalent :*

(i) There exist $\lambda_0\in C$ *and* $x_0\in H$, $x_0\neq 0$, *such that* $|\lambda_0|\geq 1$, $G_F(2\pi,0)x_0 = \lambda_0 x_0$.

(ii) There exist $\lambda_0\in C$ *and* $x_0\in H$, $x_0\neq 0$, *such that* $|\lambda_0|\geq 1$, $G_A(2\pi,0)x_0 = \lambda_0 x_0$ *and*

$C(t)G_A(t,0)x_0 = 0$ *for all* $t\geq 0$.

Moreover, if (i) or (ii) holds then :

(3.1) $$G_F(t,0)x_0 = G_A(t,0)x_0 \text{ for all } t\geq 0.$$

Proof. (i)$\Rightarrow$(ii). Let λ_0 and x_0 such that (i) holds.

Let Q_n be the mild solution to the problem

(2.2) $$\begin{cases} \dfrac{dQ_n}{dt} + A_n^*Q_n + Q_nA_n - Q_nBB^*Q_n + C^*C = 0 \\ \\ Q_n(2\pi) = Q_\#(0) \end{cases}$$

and $G_{F_n}(t,s)$ the evolution operator generated by the Yosida approximations $F_n(t)$ of $F(t)$. Since $F_n(t)$ is a continuous bounded perturbation of $A_n(t)$, it is easy to show, recalling (H3), that $\lim_{n\to\infty} G_{F_n}(t,s)x = G_A(t,s)x$ for all $x\in H$, uniformly in $t\geq s$.

By a simple computation we have

$$\frac{d}{dt}<Q_n(t)G_{F_n}(t,0)x_0,G_{F_n}(t,0)x_0> = -\|B^*Q_n(t)G_{F_n}(t,0)x_0\|^2 - \|C(t)G_{F_n}(t,0)x_0\|^2$$

from which, integrating from 0 to 2π, and letting n tend to infinity

$$(|\lambda_0|^2-1)<Q_\#(t)x_0,x_0> + \int_0^{2\pi}\{\|B^*Q_\#(t)G_F(t,0)x_0\|^2+\|C(t)G_F(t,0)x_0\|^2\}dt = 0$$

which yelds

(3.2) $\qquad B^*Q_\#(t)G_F(t,0)x_0 = 0$ for all $t\geq 0$.

(3.3) $\qquad C(t)G_F(t,0)x_0 = 0$ for all $t\geq 0$.

By (3.2) $F(t)G_F(t,0)x_0 = A(t)G_F(t,0)x_0$, this implies (3.1) and that $G_A(2\pi,0)x_0 = \lambda_0 x_0$; taking into account (3.3) the conclusion (ii) follows.

(ii)⇒(i). Let λ_0 and x_0 such that (ii) holds.

Let $Q^{(\tau)}$ and $Q_n^{(\tau)}$ be the solutions of the Riccati equation :

$$(3.4)\qquad \begin{cases} \dfrac{dQ^{(\tau)}}{dt} + A^*Q^{(\tau)}+Q^{(\tau)}A-Q^{(\tau)}BB^*Q^{(\tau)}+C^*C = 0 \\ Q^{(\tau)}(\tau) = 0 \ ,t\leq\tau \end{cases}$$

$$(3.5)\qquad \begin{cases} \dfrac{dQ_n^{(\tau)}}{dt} + A^*Q_n^{(\tau)}+Q_n^{(\tau)}A-Q_n^{(\tau)}BB^*Q_n^{(\tau)}+C^*C = 0 \\ Q_n^{(\tau)}(\tau) = 0 \ ,t\leq\tau \end{cases}$$

We recall that, by Propositions 2.1.and 2.2 we have :

$$(3.6)\qquad \lim_{\tau\to\infty} Q^{(\tau)}(t)x := Q_\#(t)x \ ; \qquad \lim_{n\to\infty} Q_n^{(\tau)}(t)x := Q^{(\tau)}x \ ; \ \forall x\in H$$

uniformly in t on bounded sets. Next we compute the derivative

$$\frac{d}{dt}<Q_n^{(\tau)}(t)G_{A_n}(t,0)x_0,G_{A_n}(t,0)x_0> = \|B^*Q_n^{(\tau)}(t)G_{A_n}(t,0)x_0\|^2 - \|C(t)G_{A_n}(t,0)x_0\|^2$$

from which, integrating from 0 to τ, and letting n tend to infinity

$$<Q^{(\tau)}(t)x_0,x_0> + \int_0^\tau \|B^*Q^{(\tau)}(t)G_A(t,0)x_0\|^2\,dt = 0$$

which implies

$$B^*Q^{(\tau)}(t)G_A(t,0)x_0 = 0, \quad \forall t\le\tau.$$

Finally, as $\tau\to\infty$ we find

$$F(t)G_A(t,0)x_0 = A(t)G_A(t,0)x_0,$$

so that $G_F(2\pi,0)x_0 = \lambda_0x_0$ and (i) holds.∎

Remark 3.2.

Assume (H1)-(H6) and that $G_F(2\pi,0)$ has only a point spectrum. By Theorem 3.1 it follows that F(t) is non resonant if one of the following conditions is fulfilled

(i) A(t) in non resonant.

(ii) A(t) is resonant but the following implication holds :

(3.7) $\quad x_0\in H,\ x_0\neq 0,\ G_A(2\pi,0)x_0 = x_0 \Rightarrow C(t)G_A(t,0)x_0 = 0$ *for at least one* $t\ge 0$.

Thus if either (i) or (ii) holds, then there exists an optimal control; otherwise an optimal control does not exists in general.

Take in fact, $H=\mathbf{R}^2$, $U = Y = \mathbf{R}$, $A(y_1,y_2) = y_2$, $C(y_1,y_2) = y_2$, $Bu = (0,u)$, $f(t) = (1,1)$. Then (H1)-(H6) hold and the state equation reduces to :

$$\begin{cases} y_1'(t) = 1 \\ y_2'(t) = u(t)+1 \end{cases}$$

so that no admissible control exists.

References.

[1] P.ACQUISTAPACE and B.TERRENI, An unified approach to abstract linear non-autonomous parabolic equations,Rend.Sem.Mat.Univ.Padova,**78**,(1987) 47-107.

[2] G.DA PRATO, Quelques résultat d'existence, unicité et régularité pour un problème de la théorie du contrôle.J. Maths. Pures Appl. **52**, (1973), 353-375.

[3] G.DA PRATO ,Synthesis of optimal control for an infinite dimensional periodic problem.SIAM J. Control & Optimiz. **25**, (1987), 706-714.

[4] G.DA PRATO, Some results on linear quadratic periodic control without detectability, Lecture Notes in Control and Information Sciences, n.114, A.Bermùdez (Editor), Springer-Verlag, Berlin (1989),100-107.

[5] G.DA PRATO and A.ICHIKAWA,Quadratic Control for Linear Periodic Systems. Appl. Math.Optim. **18**,(1988),39-66.

[6] G.DA PRATO and A.ICHIKAWA,Quadratic control for linear time varying systems. SIAM J. Control and Optimization (to appear).

[7] A.PAZY, Semigroups of Linear Operators and Applications to Partial Differential Equations, Springer-Verlag, New-York, Berlin,(1983).

[8] H.TANABE, Equations of evolution, Pitman, London (1979)

INFINITE DIMENSIONAL CONTROL PROBLEMS WITH STATE CONSTRAINTS

H. O. Fattorini
University of California, Department of Mathematics
Los Angeles, California 90024, USA

H. Frankowska
CEREMADE, Université de Paris - Dauphine
75775 Paris CX (16), France

Abstract. We study optimal control problems for distributed parameter systems with control and state constraints by using the theory of infinite dimensional nonlinear programming problems.

1. Introduction.

Optimal control problems for systems described by partial differential equations including not only control constraints but state constraints have been studied by numerous authors; see for instance [LA1], [MA1], [TR1], [RO1], [WH1]. We propose in this paper a treatment of these problems as nonlinear programming problems for functions defined in metric spaces, the constraint function taking values in a Banach space. This program has been already carried out for systems without state constraints in Hilbert spaces [FA1], [FF2], [FF1]. One attractive feature of this treatment is that (unlike, say, separation theorems) it applies to nonlinear equations and requires no convexity assumptions; other is that it is relatively simple, involving the generalization of results on the abstract nonlinear programming problem from Hilbert space valued functions to the Banach space setting, which has been done in [FR2]. Finally, the method applies as well to systems described by equations other than differential, for instance functional differential equations, with state constraints included. Details will be published elsewhere.

2. An abstract nonlinear programming problem.

Let V be a metric space, E a Banach space, Y a subset of E. Given functions $f_0 : V \to \mathbf{R} = \{\text{real numbers}\}$ and $f : V \to E$ we consider the *abstract nonlinear programming problem*

(2.1) minimize $f_o(u)$

(2.2) subject to $f(u) \in Y$.

Necessary conditions of Kuhn - Tucker type for solutions $\mathbf{u}$ of this problem, have been obtained in [FR2] (Theorem 2.1 below). We denote by $B(x, r)$ the ball of center x and radius $r \geq 0$ in an arbitrary metric space. Let $\mathbf{u} \in V$ and let $\{C(u); u \in B(\mathbf{u}, \delta)\}$ be a family of subsets of E. We denote by $\limsup_{u \to \mathbf{u}} C(u)$ (resp. $\liminf_{u \to \mathbf{u}} C(u)$) the set of all y such that $\liminf_{u \to \mathbf{u}} \operatorname{dist}(y, C(u)) = 0$ (resp. $\lim_{u \to \mathbf{u}} \operatorname{dist}(y, C(u)) = 0$.

Let Y be a subset of E, $y \in E$. The *contingent cone* $K_Y(y)$ to Y at y consists of all w in E such that there exists a sequence $\{h_k\} \subset \mathbf{R}_+ =$ {positive real numbers} with $h_k \to 0$ and a sequence $\{y_k\} \subset Y$ with $y_k \to y$ such that

$$\frac{y_k - y}{h_k} \to w \quad \text{as} \quad k \to \infty .$$

Let g be a function from V into E, $\mathbf{u} \in V$. The vector $\xi \in E$ is a (first order) *variation* of g at $\mathbf{u}$ if and only if there exists a sequence $\{h_k\} \subset \mathbf{R}_+$ with $h_k \to 0$ and a sequence $\{u_k\} \subset V$ with $d(u_k, \mathbf{u}) \leq h_k$ and such that

$$\frac{g(u_k) - g(\mathbf{u})}{h_k} \to \xi \quad \text{as} \quad k \to \infty .$$

We denote by $\partial g(\mathbf{u})$ (the *variation set* of g at $\mathbf{u}$) the set of all such ξ. Finally, Π is the canonical projection of $\mathbf{R} \times E$ into E, $(f_o, f) : V \to \mathbf{R} \times E$ is the function $(f_o, f)(u) = (f_o(u), f(u))$, and we call a set Y *sleek* at $\mathbf{y} \in Y$ if there exists $\varepsilon > 0$ such that for every $y \in Y, |y - \mathbf{y}| \leq \varepsilon$ we have

$$K_Y(y') = \liminf_{y' \to y} K_Y(y') .$$

In particular, convex sets and C^1 manifolds are sleek (see [FR2])

Theorem 2.1 Let $\mathbf{u}$ be a solution of the nonlinear programming problem (2.1) – (2.2). Assume that (a) the metric space V is complete, (b) the functions f and f_o are locally Lipschitz continuous, (c) the target set Y is closed, (d) the target set Y is sleek near $\mathbf{y} = f(\mathbf{u})$, (e) there exist constants $\varepsilon, \rho > 0$ and a compact set $Q \subseteq E$ such that for each $u \in B(\mathbf{u}, \varepsilon)$ there exists a convex, closed set $C(u) \subseteq \partial(f_o, f)(u)$, containing zerowith

(2.3) $$B(0, \rho) \subseteq \Pi(C(u)) - K_Y(y) + Q \quad (u \in B(\mathbf{u}, \varepsilon),\ y \in B(\mathbf{y}, \varepsilon) \cap K).$$

Then there exists $(z_0, z) \in \mathbf{R} \times E^*$ (E^* the dual of E) such that

(2.4) $$z_0 \geq 0, \ z \in N_Y(\mathbf{y}), \ (z_0, z) \neq 0,$$

(2.5) $$z_0\eta + \langle z, \xi \rangle \geq 0,$$

for all $(\eta, \xi) \in \liminf_{u \to \mathbf{u}} C(u)$.

See [FR2] for a proof. In the second statement (2.4), $N_Y(\mathbf{y}) \subseteq E^*$ is the normal cone to Y at $\mathbf{y}$. Condition (e) is a combined "fullness" condition on the variation sets and the target set. It is redundant in finite dimensional spaces (take $Q = B(0, \rho)$). The Hilbert space version of Theorem 2.1 can be proved under much weaker hypotheses on the functions f_0, f [FF1] [FF2]. An ancestor of this Hilbert space version (where the setup and the hypotheses are less general) was proved in [FA3]; the case where E is finite dimensional is essentially contained in the results of [EK1].

We sketch below how Theorem 2.1 can be applied to optimal control problem that include state constraints.

3. Distributed parameter systems described by elliptic differential equations.

Let Ω be a bounded domain of class $C^{(2)}$ in $\mathbf{R}^m$ with boundary Γ, and let A be a uniformly elliptic partial differential operator of class $C^{(2)}$,

$$Ay = \sum_{j=1}^{m}\sum_{k=1}^{m} \frac{\partial}{\partial x_j}\left(a_{jk}(x)\frac{\partial y}{\partial x_k}\right) + \sum_{j=1}^{m} b_j(x)\frac{\partial y}{\partial x_j} + c(x)y$$

with a boundary condition β on Γ. This boundary condition is either of Dirichlet type or of variational type $Dy = \gamma(x)y$ (D the conormal derivative). The operator A and the boundary condition β generate a strongly continuous semigroup $S(t, A, \beta)$ in the space $C(K)$ of continuous functions in K = closure of Ω, the space $C(K)$ endowed with the supremum norm (for the Dirichlet boundary condition the space $C(K)$ is replaced by its subspace $C_0(K)$ consisting of all functions vanishing on Γ).

The control system is described by the semilinear initial value problem in the space $E = C(K)$,

(3.1) $$y'(t) = \mathbf{A}(\beta)y(t) + \phi(t, y(t), u(t)),$$

(3.2) $$y(0) = y_o,$$

where $\mathbf{A}(\beta)$ is the infinitesimal generator of $S(t, A, \beta)$. Controls $u(t)$ take values a. e. in a closed, bounded subset U (called the *control set*) of a Banach space F. The nonlinear term $\phi(t, y, u)$ is defined, continuous and locally bounded in $[0, T] \times C(K) \times U$ and takes values in $L^\infty(\Omega)$; we assume that it possesses a Fréchet derivative $\partial_y\phi(t, y, u)$ with respect to y which is strongly continuous and locally bounded. Admissible controls are assumed to be in a space of F – valued measurable functions, where the notion of measurability must be such that $t \to \phi(t, y(t), u(t))$ is $(L^1(\Omega), L^\infty(\Omega))$ – weakly measurable for every admissible control. For instance, if $F = L^\infty(\Omega)$ and $\phi(t, y, u) = \psi(t, y) + u$, the space of admissible controls consists of all $(L^1(\Omega), L^\infty(\Omega))$ – weakly measurable $L^\infty(\Omega)$ - valued functions $u(\cdot)$ such that $u(t) \in U$ a. e. The treatment of (3.1) – (3.2) under these measurability assumptions is slightly nonstandard (see [FA2] for the linear case) but existence and uniqueness of $C(K)$ – valued solutions $y(t) = y(t, u)$ in intervals $0 \le t \le T'$, is proved by successive approximations in the usual way. We note that in general $T' < T$, that is, solutions may blow up somewhere in the interval $0 \le t \le T$.

We use a cost functional of the form

$$y_o(t, u) = \int_0^t \phi_o(s, y(s, u), u(s))ds$$

where the assumptions on ϕ_o are similar to those on ϕ; $\phi_o(t, y, u)$ is defined, continuous and locally bounded in $[0, T] \times C(K) \times U$ and possesses a Fréchet derivative $\partial_y\phi_o(t, y, u)$ with respect to y which is continuous and locally bounded as a E^* - valued function. Finally, we assume that $t \to \phi_o(t, y(t), u(t))$ is measurable for every admissible control. The problem is

(3.3) $$\text{minimize } y_o(\bar{t}, u)$$

(3.4) $$\text{subject to } y(\bar{t}, u) \in Y,\ y(t, u) \in X(t)\ (0 \le t \le \bar{t}).$$

The *target set* Y and the variable *constraint set* $X(t)$ are subsets of $C(K)$. Assuming that the endpoint $\bar{t}$ of the control interval $0 \le t \le \bar{t}$ is fixed we can simply replace $X(\bar{t})$ by $X(\bar{t}) \cap Y$ and reduce the target condition and the state constraints to the unique state constraint

$$y(t, u) \in X(t) \quad (0 \le t \le t).$$

However, it is more convenient to keep the target condition and the state constraint separated as in (3.4).

We can mold the problem (3.3) – (3.4) into a nonlinear programming problem (2.1) – (2.2) as follows. The space V is the space of all admissible controls with the distance

(3.5) $$d(u, v) = \text{meas}\{t;\ u(t) \neq v(t)\}\ .$$

(or rather a ball $B(\mathbf{u}, \delta) \subset V$, see Lemma 3.1 below). The space E is the Cartesian product $C(K) \times C([0, \mathbf{t}]; C(K)) = C(K) \times C(([0, \mathbf{t}] \times K)$ endowed with its usual supremum norm. Assuming that $\mathbf{t}$ is fixed, the functions $f_0(u)$ and $f(u)$ are defined by

(3.6) $$f_0(u) = y_0(\mathbf{t}, u)\ ,$$

(3.7) $$f(u) = (y(\mathbf{t}, u), y(\cdot\,, u))\ ,$$

where $y(t, u)$ is the solution of (3.1) corresponding to the control $u \in V$. Since $y(t, u)$ may not exist in the whole control interval $0 \le t \le \mathbf{t}$, we may either impose conditions on global existence or use the following result:

Lemma 3.1 Let $\mathbf{u} \in V$ be a control such that $y(t, \mathbf{u})$ exists in $0 \le t \le \mathbf{t}$. Then there exists $\delta > 0$ such that, if $d(u, \mathbf{u}) \le \delta$ (d the distance (3.5)) then $y(t, u)$ exists in $0 \le t \le \mathbf{t}$ and the map $u \to y(t, u)$ is Lipschitz continuous in $B(\mathbf{u}, \delta)$ uniformly in $0 \le t \le \mathbf{t}$.

Lemma 3.1 is applied to the optimal control $\mathbf{u}$, assumed to exist; by definition it produces a global trajectory, thus trajectories corresponding to neighboring controls will be global as well.

The target set $\mathbf{Y}$ for the nonlinear programming problem under construction is

(3.8) $$\mathbf{Y} = \{(y, y(\cdot)) \in C(K) \times C([0, \mathbf{t}]; C(K));\ y \in Y, y(t) \in X(t)\ \ (0 \le t \le \mathbf{t})\} =$$

$$= Y \times \{y(\cdot) \in C([0, \mathbf{t}]; C(K));\ y(t) \in X(t)\ \ (0 \le t \le \mathbf{t})\} = Y \times Z\ .$$

We compute variations of the function (3.7) at an arbitrary control $u \in V$ using an extension of Theorem 4 in [LY1].

Lemma 3.2 Let $u(\cdot)$, $v(\cdot)$ be admissible controls, $0 < \rho \le 1$. Then there exists a measurable set $e_\rho \subseteq [0, t]$ with $\mathrm{meas}(e_\rho) \le \rho$ and such that if u_ρ is the control defined by

$$u_\rho(t) = v(t) \quad (t \in e_\rho), \; u_\rho(t) = u(t) \quad (t \notin e_\rho),$$

then the solution $\xi(t)$ of the linearized initial value problem

$$\xi'(t) = \{A(\beta) + \partial_y\phi(t, y(t, u), u(t))\}\xi(t) + \tag{3.9}$$

$$+ \{\phi(s, y(s, u), v(s)) - \phi(s, y(s, u), u(s))\} \quad (0 \le t \le t),$$

$$\xi(0) = 0 \tag{3.10}$$

satisfies the asymptotic relation

$$y(t, u_\rho) = y(t, u) + \rho\xi(t) + o(\rho) \quad \text{as} \quad \rho \to 0+ ,$$

It follows from this result that $(\xi(t), \xi(\cdot)) \in \partial f(u)$. The variations of f_o are similarly computed; we have

$$y_o(t, u_\rho) = y_o(t, u) + \rho\xi_o(t) + o(\rho) \quad \text{as} \quad \rho \to 0+$$

where the function $\xi_o(t)$ is given by

$$\xi_o(t) = \int_0^t \{\phi_o(s, y(s, u), v(s)) - \phi_o(s, y(s, u), u(s))\} ds + \tag{3.11}$$

$$+ \int_0^t \langle \partial_y\phi_o(s, y(s, u), u(s)), \xi(s) \rangle ds ,$$

and $\xi(t)$ is the solution of (3.8) – (3.9).

Taking limits, we deduce that any vector of the form $(\xi_o(t), \xi(t), \xi(\cdot))$ in the space $\mathbf{R} \times C(K) \times C([0, t]; C(K))$, where $\xi(\cdot)$ is the uniform limit of solutions of the initial value problem (3.9) – (3.10) and $\xi_o(t)$ is the uniform limit of (3.11) belongs to $\partial(f_o, f)(u)$. It follows then from [FR1] (or from extension of the results in [FA1], since the equation is linear) that $\partial(f_o, f)(u)$ contains any vector of the form $(\xi_o(t), \xi(t), \xi(\cdot))$ where $\xi(\cdot)$ is the solution of the initial value problem

(3.12) $$\xi'(t) = \{\mathbf{A}(\beta) + \partial_y\phi(t, y(t, u), u(t))\}\xi(t) + v(t) \quad (0 \le t \le \bar{t})$$

(3.13) $$\xi(0) = 0$$

and $\xi_o(t)$ is given by

$$\xi_o(t) = \int_0^t v_o(s)\,ds + \int_0^t \langle \partial_y\phi_o(s, y(s, u), u(s)), \xi(s)\rangle\, ds\,,$$

with $(v_o(t), v(t)) \in \mathbf{conv}\{(\phi_o, \phi)(t, y(t, u), U) - (\phi_o, \phi)(t, y(t, u), u(t))\}$ a. e. (**conv** means closed convex hull).

We apply Theorem 2.1. The dual space $E^* = (C(K) \times C([0, \bar{t}] \times K))^*$ can be identified with the space $\Sigma(K) \times \Sigma([0, \bar{t}] \times K))$ of all pairs $(\lambda(dx), \mu(dtdx))$, where $\lambda(dx)$ is a finite regular Borel measure in K and $\mu(dtdx)$ is a finite regular Borel measure in $[0, \bar{t}] \times K$; this space is endowed with the total variation norm

$$\|(\lambda, \mu)\| = \int_K |\lambda(dx)| + \int_0^{\bar{t}}\int_K |\mu(dxdt)|\,.$$

The key point is the verification of (2.3) for the target set Y. Denote by $C(u)$ the set of variations $(\xi_o(\bar{t}), \xi(\bar{t}), \xi(\cdot))$ constructed above, and consider the set $\Pi(C(u)) = \{(\xi(\bar{t}), \xi(\cdot))\}$. In general, $\{\xi(\cdot)\}$ cannot contain an interior point in $C([0, \bar{t}]; C(K))$, for this would mean that every function in a neighborhood in $C([0, \bar{t}]; C(K))$ could be impersonated by a trajectory of the system, which contradicts smoothing properties of parabolic equations. The same properties imply that the set $\{\xi(\bar{t})\}$ cannot contain interior points in $C(K)$. Now, since Y is sleek, for any $(y, y(\cdot)) \in C(K) \times C(([0, \bar{t}] \times K)$ we have

$$K_Y(y, y(\cdot)) = K_Y(y) \times K_Z(y(\cdot))\,.$$

Thus, we can insure (2.3) by assuming that the $X(t)$ are, say, convex and contain a common open set, and requiring that the target set Y satisfy: there exists $\rho > 0$ and a compact set Q such that

(3.14) $$B(0, \rho) \subseteq K_Y(y) + Q \quad (y \in B(\bar{y}, \varepsilon))\,.$$

This (and the sleekness condition) is satisfied, for instance, if Y is convex with interior points or if Y is a C^1 manifold of finite codimension.

Replacing $C(\mathbf{u})$ by its closure, the Kuhn - Tucker inequality (2.4) for $(z_o, z) = (z_o, \lambda(dx), \mu(dtdx)) \neq 0$ is

$$z_o\xi_o(\mathbf{t}) + \int_K \xi(\mathbf{t}, x)\lambda(dx) + \int_0^{\mathbf{t}}\int_K \xi(t, x)\mu(dtdx) \geq 0 ,$$

to be satisfied for all $(\eta, \xi) \in \liminf_{u \to \mathbf{u}} C(u)$. The variations $(\xi_o(\mathbf{t}), \xi(\mathbf{t}), \xi(\cdot))$ given by (3.9), (3.10) and (3.11) depend continuously on the admissible control $u(\cdot)$ in the distance of V, thus the vectors $(\xi_o(\mathbf{t}), \xi(\mathbf{t}), \xi(\cdot))$ corresponding to $u = \mathbf{u}$ belong to $\liminf_{u \to \mathbf{u}} C(u)$. We exploit the arbitrariness of $v(\cdot)$ by using in its place spike perturbations $\mathbf{u}_{r,s,v}(\cdot)$ of $\mathbf{u}(\cdot)$,

$$\mathbf{u}_{r,s,v}(t) = v \quad (s - r \leq t \leq s), \quad \mathbf{u}_{r,s,v}(t) = \mathbf{u}(t) \quad \text{elsewhere},$$

where v is an arbitrary element of the control set U (see [FA3]) and then letting $r \to 0+$. This produces the limit functions

$$\xi(t) = 0 \quad (0 \leq t < s) ,$$

$$\xi(t) = S(t, s; \mathbf{u})\{\phi(s, y(s, \mathbf{u}), v) - \phi(s, y(s, \mathbf{u}), \mathbf{u}(s))\} \quad (s \leq t \leq \mathbf{t}) ,$$

where the operator $S(t, s; \mathbf{u})$ is the solution operator of the linear equation

$$\xi'(t) = \{A(\beta) + \partial_y\phi(t, y(t, u), u(t))\}\xi(t)$$

and

$$\xi_o(t) = 0 \quad (0 \leq t < s) ,$$

$$\xi_o(t) = \phi_o(s, y(s, \mathbf{u}), v) - \phi_o(s, y(s, \mathbf{u}), \mathbf{u}(s)) +$$

$$+ \int_s^t \langle \partial_y\phi_o(\sigma, y(\sigma, u), \mathbf{u}(\sigma)), \xi(\sigma)\rangle \, d\sigma \quad (s \leq t \leq \mathbf{t}) .$$

Manipulations similar to those in [FA3] produce then the maximum (or, rather, minimum) principle

$$(3.15)\qquad z_o\phi_o(t, y(t, \mathbf{u}), \mathbf{u}(t)) + \langle z(t), \phi(t, y(t, \mathbf{u}), \mathbf{u}(t))\rangle \le$$

$$\le z_o\phi_o(t, y(t, \mathbf{u}), v) + \langle z(t), \phi(t, y(t, \mathbf{u}), v)\rangle \quad (v \in U)$$

a.e. in the control interval $0 \le t \le \mathbf{t}$, where $z(t)$ is the solution of the backwards initial value (or "final value") problem

$$(3.16)\qquad dz(t) = -\{\mathbf{A}(\beta) + \partial_y\phi(t, y(t, \mathbf{u}), \mathbf{u}(t))\}^* z(t)dt -$$

$$-\partial_y\phi_o(t, y(t, \mathbf{u}), \mathbf{u}(t))dt - \mu(dt) \qquad (0 \le t \le \mathbf{t})$$

$$(3.17)\qquad z(\mathbf{t}) = \lambda$$

in the space $\Sigma(\Omega)$, where $\mu(dt)$ is the measure $\mu(dtdx)$ considered as a $\Sigma(\Omega)$ - valued measure.

We note that the case where the equation is linear, the constraint set $X(t)$ is convex and has a nonempty interior and Y is an arbitrary convex closed set (possibly without interior points) can be treated using the separation theorem for convex sets in Banach spaces along the lines of [FA2].

4. Distributed parameter systems described by hyperbolic differential equations.

The abstract nonlinear programming formulation in §2 apply as well to systems described by semilinear hyperbolic initial value problems

$$(4.1)\qquad y''(t) = \mathbf{A}(\beta) + \phi(t, y(t), u(t)),$$

$$(4.2)\qquad y(0) = y^o, \; y'(0) = y^1,$$

where now $\mathbf{A}(\beta)$ is the (cosine function generator) corresponding to A and the boundary condition β in the space $L^2(\Omega)$. The treatment can handle, for instance, constraints of the form

$$(4.3)\qquad E(y(t)) \le C \;\; (0 \le t \le \mathbf{t}),$$

where $E(y(t))$ is the energy of the solution at time t. (its norm in the space H below). The equation (4.1) is reduced to a first order system for the function $(y(\cdot), y'(\cdot))$ in the energy space $H = H^1(\Omega) \times L^2(\Omega)$. ($H^1{}_{,o}(\Omega)$ in the case of the Dirichlet boundary condition; see [FA4] for details). The problem is

(4.4) minimize $y_o(t, u)$

(4.5) subject to $(y(t, u), y'(t, u)) \in Y$, $(y(t, u). y'(t, u)) \in X(t)$ $(0 \leq t \leq t)$,

where $y_o(t, u)$ is a suitable cost functional and the $X(t)$ are sets in the energy space H. The space V is defined in the same way as in §3. The space E is the Cartesian product $H \times C_w([0, t] ; H)$, where $C_w([0, t] ; H)$ is the space of all weakly continuous H - valued functions defined in $0 \leq t \leq t$ endowed with the supremum norm. The dual of this space can be expressed in terms of a complete orthonormal system $\{e_n\}$ in H; $C_w([0, t] ; H)^*$ is the space $\Sigma_w([0, t] ; H)$ of all sequences $\{\mu_n\}$ of measures $\mu_n \in \Sigma([0, t])$ such that

$$\sum_{n=1}^{\infty} \int_0^t \langle e_n, f(s) \rangle \mu_n(ds) \leq C|f|$$

for $f \in C_w([0, t] ; H)$. The functions $f_o(u)$ and $f(u)$ are defined by

(4.6) $f_o(u) = y_o(t, u)$

(4.7) $f(u) = (y(t, u), y'(t, u), y(\cdot, u), y'(\cdot, u))$

where t is the endpoint of the control interval $0 \leq t \leq t$ and $y(t, u)$ is the solution of (4.1) – (4.2) corresponding to the control $u \in V$. The considerations in §3 about global existence apply. The target set Y is

$$\{(y^o, y^1, y^o(\cdot), y^1(\cdot)) \in H \times C_w([0, t]; H); (y^o, y^1) \in Y, (y^o(t), y^1(t)) \in X(t) \ (0 \leq t \leq t)\} .$$

The rest of the treatment is similar to that of the parabolic problem in §3. We note that, due to the favorable controllability properties of the hyperbolic equation (4.1) (see [FA2]) it is possible to handle point targets $Y = \{y^o, y^1\}$.

We note another important difference between the treatment of parabolic problems and that of hyperbolic problems; in the former, pointwise constraints can be handled (this is implicit in the $C(K)$ setting) whereas in the latter, the results are restricted to constraints like (4.3) of integral type on the state variables: pointwise constrains such as those in [WH1] are not included.

We point out that the abstract nonlinear programming setting applies to many other different situations. For instance, other parameters in the initial value problems (3.1) – (3.2) or (4.1) – (4.2) (for instance, the initial conditions) can be considered as controls. Also, we may treat optimal control problems for functional differential equations with target conditions of functional type, including state constraints: all that needs to be done is to compute the variations of the maps $f(u)$, $f_0(u)$ corresponding to the equation and to the cost functional. This computation was carried out in a particular case in [FA3].

The work of the first author was supported in part by the National Science Foundation under grant DMS - 8701877

References.

[EK1] I. EKELAND, Nonconvex minimization problems, Bull. Amer. Math. Soc. N. S. 3 (1979) 443-474

[FA1] H. O. FATTORINI, A remark on the "bang - bang" principle for linear control systems in infinite dimensional spaces, SIAM J. Control 6 (1968) 109-113

[FA2] H. O. FATTORINI, The time optimal control problem in Banach spaces, Appl. Math. Optimization 1 (1974) 163-188.

[FA3] H. O. FATTORINI, A unified theory of necessary conditions for nonlinear nonconvex control systems, Appl. Math. Optim. 15 (1987) 141-185.

[FA4] H. O. FATTORINI, Optimal control of nonlinear systems: convergence of suboptimal controls, II, Springer Lecture Notes on Control and Information Sciences 97 (1987) 230-246

[FF1] H. O. FATTORINI and H. FRANKOWSKA, Necessary conditions for infinite dimensional control problems, Proceedings of 8th. International Conference on Analysis and Optimization of Systems, Antibes - Juan-Les-Pins, 1988. To appear in Lecture Notes on Control and Information Sciences.

[FF2] H. O. FATTORINI AND H. FRANKOWSKA, Necessary conditions for infinite dimensional control problems, to appear in Mathematics of Control, Signals and Systems.

[FR1] H. FRANKOWSKA, Estimations a priori pour les inclusions différentielles opérationnelles, C. R. Acad. Sci. Paris 308 Série I (1989) 47-50.

[FR2] H. FRANKOWSKA, Some inverse mapping theorems, to appear in Annales de IHP, Analyse Nonlineaire, (1990).

[LA1] I. LASIECKA, State constrained control problems for parabolic systems: regularity of optimal solutions, Appl. Math. Optimization 6 (1980) 1-29.

[LY1] LI X. and YAO Y., Maximum principle of distributed parameter systems with time lags, *Distributed Parameter Systems*, Springer Lecture Notes in Control and Information Sciences 75 (1985) 410-127.

[MA1] U. MACKENROTH, On some elliptical control problems with state constraints, Optimization 17 (1986) 595-607.

[RO1] T. ROUBICEK, Generalized solutions of constrained optimization problems, SIAM J. Control and Optimization 24 (1986) 951-960.

[TR1] F. TRÖLTZSCH, On some parabolic boundary control problems with constraints on the control and functional constraints on the state, Zeitschrift für Analysis und Ihre Anwend. 1 (1982) 1-13.

[WH1] L. W. WHITE, Control of a hyperbolic problem with pointwise stress constraints, JOTA 41 (1983) 359-369.

APPROXIMATE AND NUMERICAL METHODS OF THE OPTIMAL CONTROL SYNTHESIS FOR STOCHASTIC SYSTEMS

Vladimir B. Kolmanovskii and Gennadii Ye. Kolosov
Moscow Institute of Electronic Machinery
109028, Moscow, USSR

This paper is a survey of some results connected with the numerical and approximate synthesis of the optimal control for stochastic dynamical systems with concentrated or distributed parameters. Approximate methods of the solution of optimal control stochastic problems have double interest. First it's well known that exact analytical solution of such problems may be obtained only in the exclusive cases. So usually approximate methods give the possibility to obtain the efficient way of determining the control algorithm close to the optimal one. On the other hand according to the dynamic programming method many optimal control problems may be reduced to the solution of some special nonlinear equations with partial or functional derivatives (the Bellman equations).Thus for the problem (1.1),(1.2) considered below the corresponding equation (1.3) is a nonlinear equation of parabolic type. The solution of the latter equation may be of interest for the theory of the appropriate systems with distributed parameters. In this case the methods and ideas of the optimal control approximate synthesis may be useful for the approximate solutions construction and qualitative theory of the equations like (1.3).

The algorithms considered below are founded either on some small parameters in the system equations or on successive approximations procedure for the Bellman equation solution. Numerical methods which are effective as a rule for the systems of small dimension are illustrated for some concrete examples.

Bibliography does not pretend on the fullness. It contains only the sources which were essentially used in this survey, but at the same time which contain further extensive information on the problems under consideration.

1.Approximate synthesis of the optimal control for the systems with small parameters. First let us set forth formally the methods of approximate synthesis for dynamical systems described by a vector-valued stochastic differential equation of the Ito type

$$dx(t) = f(t,x,u)dt + \sigma(t,x,u)d\xi(t), \quad 0 \le t \le T, \quad x(0) = x_0. \qquad (1.1)$$

Here $x \in R_n$ is phase vector, $u \in U \subset R_r$ is a control, $\xi(t) \in R_1$ denotes a standard Wiener process and R_n is an Euclidean space of dimension n. The matrices f and σ of the corresponding dimensions, the time moment T, the vector x_0 and the set U are given. The performance index subject to minimization is defined by the expression

$$I(u) = M\,[F_1(x(T)) + \int_0^T F_2(t,x(t),u)dt] \rightarrow \inf_{u\in U} \tag{1.2}$$

where F_1, F_2 are given scalar penalty functions. M is the expectation. Under the assumption that the current values of the vector $x(t)$ can be exactly measured it is required to find the optimal control u_0 in the form of synthesis $u_0 = u_0(t,x(t))$. Let us denote by $V(t,x)$, $0 \le t \le T$, $x \in R_n$ the Bellman function of the stated problem. Then $I(u_0) = V(0,x_0)$. Under some assumptions according to dynamic programming method function V is a solution of the following Cauchy problem

$$\inf_{u\in U} [L_u V(t,x) + F_2(t,x,u)] = 0, \quad V(T,x) = F_1(x), \tag{1.3}$$

$$L_u = \frac{\partial}{\partial t} + f'(t,x,u)\frac{\partial}{\partial x} + \frac{1}{2}\mathrm{Tr}_1\sigma_1(t,x,u)\frac{\partial^2}{\partial x^2}, \quad \sigma_1 = \sigma\sigma',$$

where prime denotes transposition and Tr is the matrix trace.

It should be stressed that the greatest lower bound in equation (1.3) must by calculated with respect to the vector parameter $u \in U$. Therefore the dynamic programming method leads to the following algorithm of the solution of the optimal control problem: first of all it is necessary to find the function $V(t,x)$ by solving the problem (1.3) and after that to obtain the optimal control u_0 by solving the finite dimensional optimization problem

$$L_u V(t,x) + F_2(t,x,u) \rightarrow \inf_{u\in U} \tag{1.4}$$

The obtained function $u_0 = u_0(t,x)$, depends on (t,x) and consequently is the control synthesis. However, exact analytical solutions for the Eqs. of (1.3) type can be found only in exceptional cases, for instance in linear-square and some scalar problems.

That is why different approximate and numerical methods of solving Eq.(1.3) (hence the synthesis problem) acquire great practical importance. The efficiency of approximate synthesis methods, as a rule, is brought about by the presence of a small parameter in the problem (1.1), (1.2). Suppose that functions f, σ, F_1, F_2 in (1.1), (1.2) depend on parameter ε

$$f = f(t,x,u,\varepsilon), \quad \sigma = \sigma(t,x,u,\varepsilon), \quad F_2 = F_2(t,x,u,\varepsilon), \quad F_1 = F_1(x,\varepsilon) \tag{1.5}$$

Represent V(t,x) in form

$$V = V_0(t,x) + \varepsilon V_1(t,x) + \dots \tag{1.6}$$

Then substitute (1.6) in (1.3) and expand the left-hand side of (1.3) and the function F_1 into a series of powers of ε. Then equating to zero the expressions for different powers of ε we obtain a system of equations for functions $V_i(t,x)$ in (1.6). Suppose that for some value of j the indicated equations are solved for all $i=0,1,\dots,j$ and the functions $V_i(t,x)$, $i \le j$ are found. Then to determine the j-th approximation to optimal control it is necessary to solve the problem (1.4) whose left-hand side is expanded in ε taking into consideration (1.5),(1.6) and expansion terms up to the order j. The control being obtained in such a way will be denoted by $v_j(t,x)$. In some situations it is possible to prove that

$$\begin{aligned} &V(t,x) - \sum_{i=0}^{j} \varepsilon^i V_i(t,x) = O(\varepsilon^{j+1}) \\ &I(u_0) - I(v_j) = O(\varepsilon^{j+1}). \end{aligned} \tag{1.7}$$

There are two ways for the proof of the relations (1.7). In the first place using (1.3) and the equations for V_i one estimates the differences

$$V - \sum_{i=0}^{j} \varepsilon^i V_i \quad \text{and} \quad I(v_j) - \sum_{i=0}^{j} \varepsilon^i V_i .$$

The different way is founded on the direct analysis of the equations (1.1) and the cost functional (1.2). Consider some concrete cases in detail.

1.1. Small stochastic disturbances. Let the functions f, F_1 and F_2 don't depend on ε and the matrix σ be equal to $\sqrt{\varepsilon}\,\sigma(t,x)$. The problem (1.3) takes the form

$$\begin{aligned} &V_t + H(t,x,V_x) + \frac{\varepsilon}{2} \operatorname{Tr} \sigma_1 V_{xx} = 0, \qquad V(T,x) = F_1(x), \\ &H(t,x,V_x) = \inf_{u \in U} \Big[f'(t,x,u)V_x + F_2(t,x,u) \Big], \; V_t = \frac{\partial V}{\partial t}, \; V_x = \frac{\partial V}{\partial x} . \end{aligned} \tag{1.8}$$

The solution of the problem (1.8) will be found in the form (1.6). Substitute (1.6) in (1.8) and equate to zero the coefficients for the same degrees of ε. Then we get the relations that define functions V_i. According to this scheme equation of the i-th approximation turns out to be linear in V_i for $i \ge 1$. In particular we get for $i=0,1$

$$V_{0t} + H(t,x,V_{0x}) = 0, \quad V\ (T,x) = F_1(x), \tag{1.9}$$

$$V_{1t} + \frac{\partial H(t,x,V_{0x})'}{\partial V_x} V_{1x} + \frac{1}{2} \operatorname{Tr} \sigma_1 V_{0xx} = 0, \; V_1(T,x) = 0. \tag{1.10}$$

The control of the i-th approximation $v_i(t,x)$according to (1.4) is defined by the equality

$$\inf_{u\in U}[f'(t,x,u)\frac{\partial}{\partial x}(V_0+\ldots+\varepsilon^i V_i) + F_2(t,x,u)]$$
$$=f'(t,x,v_i)\frac{\partial}{\partial x}(V_0+\ldots+\varepsilon^i V_i) + F_2(t,x,v_i).$$

It's obvious from here and (1.9) that V_0 is a Bellman function and v_0 is optimal control in the determined problem (1.1),(1,2) with $\sigma=0$. The consequent approximations $V_i(t,x)$ are determined by quadrature from a some function of the preceding approximations along the trajectories of the system (1.1) with $\sigma = 0$ and control $u = v_0$. For example by virtue of (1.10) under some conditions

$$V_1(t,x) = \frac{1}{2} \int_t^T \operatorname{Tr} \sigma_1(\tau,y(\tau))V_{0xx}(\tau,y(\tau))d\tau$$

Here $y(\tau)$ is a solution of Eq. (1.1) for $\sigma \equiv 0$, $u \equiv v_0$ on the segment $t \le \tau \le T$ with initial condition $y(t) = x$. The proof of the relation (1.7) for the case of small disturbances under consideration is contained in the papers [20,24].

1.2. Quasi linear systems. Let the problem (1.1),(1.2) has the form

$$dx(t)=(B(t)u + Ax + \varepsilon f(t,x))dt + \sigma(t)d\xi(t), \; 0\le t\le T, \; x(0)=x_0$$

$$J(u)=M[x'(T)N_3x(T) + \int_t^T (x'(t)N_1(t)x(t) + u'N_2u)dt \tag{1.11}$$

It's assumed that $U=R_r$, matrices N_i are bounded nonnegative definite and N_2 is positive definite. Function $V_0(t.x)$ in the representation (1.6) equals

$$V_0 = x'P(t)x + \int_t^T \operatorname{Tr} \sigma_1 P(s)ds. \tag{1.12}$$

Here matrix $P(t)$ is defined by the relations

$$\dot{P}(t) + A'P + PA - PB_1P + N_1 = 0, \; P(T) = N_3, \; B_1 = BN_2^{-1}B'. \tag{1.13}$$

The approximations V_j, $j\ge 1$ satisfy to the linear equations

$$V_{jt} + \frac{1}{2} \operatorname{Tr} \sigma_1 V_{jxx} + f'V_{(j-1)x} - \frac{1}{4} \sum_{i=0}^{j} V_{ix}' B_1 V_{(j-i)x} = 0,$$
$$V_j(T,x) = 0. \tag{1.14}$$

According to (1.4),(1.6),(1.11) the control of j-th approximation v (t,x) is given by the formula

$$v_j(t,x) = -\frac{1}{2} N_2^{-1}B'(V_{0x}+\dots+\varepsilon^j V_{jx}) \tag{1.15}$$

The solution of equation (1.14) has the form

$$V_j(t,x) = M\int_t^T \Big[f'(s,x)V_{(j-1)x}(s,x) - \frac{1}{4}\sum_{i=1}^{j-1} V'_{ix}(s,y)B_1(s)V_{(j-i)x}(s,y)\Big]ds \tag{1.16}$$

Where $y = y(s)$ is the solution of the equation (1.11) with $\varepsilon = 0$, $u = v_0$ and initial condition $y(t) = x$.

The proof of the estimations (1.7) for quasi linear systems was given in [4, 23-25,29].

1.3. Adaptive systems with small a priori uncertainty. Consider the problem (1.11) for $f \equiv 0$, $\sigma_1 = \sigma\sigma' > 0$ where elements of a constant matrix A are a priori unknown and represent some Gaussian vector from R_{n^2} with expectation $m_0 \in R_{n^2}$ and co variance matrix εD_0 of dimension $n^2 \times n^2$. In this case for the description of the controlled system dynamics it's necessary to add to equation (1.11) with $f \equiv 0$ the Kalman filtering equations for vector $m(t)$ and matrix $D(t)$ which characterize the a posteriori density of the conditional probability distribution of the matrix A elements. These equations have the form

$$dm(t) = DR'\sigma_1^{-1}\Big[dx(t) - (\bar{A}(m)x + Bu)dt\Big], \quad m(0) = m_0,$$

$$\dot{D} = -DR_1D, \quad D(0) = \varepsilon D_0, \quad R_1 = R'\sigma_1^{-1}R.$$

Here matrix $\bar{A}(m)$ is obtained from A by the change of the unknown elements of A by their aposteriori mean value m. The matrix R(x) of dimension $n \times n^2$ equals

$$R(x) = \begin{bmatrix} x_1 \dots x_n & 0 \dots 0 & \dots & \dots & 0 \dots 0 \\ 0 \dots 0 & x_1 \dots x_n & \dots & \dots & 0 \dots 0 \\ \dots \dots & \dots \dots & \dots & \dots & \dots \dots \\ 0 \dots 0 & 0 \dots 0 & \dots & \dots & x_1 \dots x_n \end{bmatrix}.$$

Bellman function $V(t,x,m,D)$ satisfies to the relations

$$V_t + x'A'(m)V_x + \inf_{u \in R_r}(u'B'V_x + u'N_2u) + x'N_1x$$
$$+ \frac{1}{2}\operatorname{Tr}\Big[DR_1D(V_{mm} - 2V_D) + 2DR'V_{xm} + \sigma_1 V_{xx}\Big] = 0, \quad V(T,x,m,D) = N_3. \tag{1.17}$$

Let us find the solution of problem (1.17) like (1.6) in the form

$$V(t,x,m,D) = V_0(t,x,m) + \varepsilon V_1(t,x,m,D) + \dots \tag{1.18}$$

The control of the j-th approximation $v_j(t,m,x,D)$ is given by (1.15). Note that in (1.18) function V_0 does not depend on D and is defined by

(1.12),(1.13).A in (1.13) being replaced by matrix $\bar{A}(m)$ with the fixed value of m (just the same as in the left side (1.18)). Functions V_j, $j \geq 1$ in (1.18) satisfy to the relations

$$V_{jt} + x'(\bar{A}(m) - PB_1)V_{jx} + \frac{1}{2}\mathrm{Tr}\sigma_1 V_{jxx} + \alpha_j(t,x,m,D) = 0, V_j(T,x,m,D) = 0. \quad (1.19)$$

Here $$\alpha_j(t,x,m,D) = \frac{1}{2}\mathrm{Tr}\Big[DR_1D(V_{(j-2)mm} - 2V_{(j-1)D} + 2DR'V_{(j-1)xm}\Big]$$

$$-\frac{1}{4}\sum_{i=1}^{j-1} V'_{ix}B_1V_{(j-i)x}, \; j\geq 2, \; \alpha_1 = \mathrm{Tr}\Big[DR'(x)V_{0xm}(t,x,m)\Big]. \quad (1.20)$$

The problem (1.19) solution may be represented in the form

$$V_j(T,x,m,D) = \int_t^T \alpha_j(s,y,m,D)ds. \quad (1.21)$$

Where $y = y(s)$ is the solution of the equation

$$dy(s) = \Big[\bar{A}(m) - B_1(s)\Big]y(s) + \sigma(s)d\xi(s), \; t \leq s \leq T, \; y(t) = x. \quad (1.22)$$

Remark that x,m and D in formulae (1.21),(1.22) must be considered as constant. The error estimate of the described method of approximate synthesis was established in [8].Adaptive hereditary systems with small a priori uncertainty were investigated in [7].

Remark 1. The methods of approximate synthesis stated above were devoted to the regular disturbed systems. In singular disturbed systems the phase vector has two group of components - fast y(t) and slow z(t) which are described by the equations

$$dz(t) = f_1(t,z,y,u)dt + \sigma_1(t,z,y,u)d\xi_1(t)$$
$$\varepsilon dy(t) = f_2(t,z,y,u)dt + \sqrt{\varepsilon}\,\sigma_2(t,z,y.u)d\xi_2(t)$$

Here ξ_i are standard Wiener processes. Besides singular perturbed problems arise in systems with "cheap" control. As examples we may indicate the linear-quadratic problem (1.10) with $f=0$, $N_2 = \varepsilon\bar{N}_2$ and problem of the optimal estimation with small noises in the observation channel [2]. The approximate synthesis problems of optimal control for singular perturbed stochastic systems were investigated in [26,27,33-35,38]. The method of integral manifolds was applied in [18] for the investigation of singular perturbed Riccatti equations.This gives the possibility to calculate a fast component of the solution only with the aid of algebraic operations.

Remark 2. The above considered methods were devoted only to approximate synthesis of the optimal control. But if the phase vector cannot be measured then the same problems statements are valid for the open-loop optimal control. Quasioptimal open-loop control was derived in [32] for linear systems with small stochastic disturbances by reduction a stochastic problem to determined one. Stochastic maximum principle was used in [3] for the approximation of the optimal open-loop control of quasi linear systems.

Remark 3. The principle of the generalized work [12] is effective for the construction of approximate optimal control in some systems.

2. Successive approximations. Successive approximations method is founded on the approximation of the Bellman equation solution by a sequence of solutions of some linear equations.

2.1. Bellman method of successive approximation [1] for the problem (1.3). Take an arbitrary admissible control v_0 and substitute it in the (1.3) left side instead of u. After that find the solution V_0 of the corresponding linear problem. Substitute V_0 in (1.4) and determine a control u_1 that minimizes the (1.4) left side. Continue this procedure. As a result we get the sequences $V_i(t,x)$ and $v_i(t,x)$ such that

$$L_{v_i}V_i + F_2(t,x,v_i) = 0, \quad V_i(T,x) = F_1(x),$$
$$\inf_{u\in U} [L_uV_i + F_2(t,x,u)] = L_{v_{i+1}}V_i + F_2(t,x,v_{i+1}) \tag{2.1}$$

It's proved that under some conditions [21,24,31] sequence V_i converges to the solution of the problem (1.3) and sequence $v_i(t,x)$ converges to the optimal control. In particular for linear-quadratic problem (1.11) with $f = 0$, $v_0 = 0$ procedure (2.1) leads to approximation of the solution of the Riccati equation (1.13) by the sequence P_i satisfying to linear equations

$$\dot{P}_i + A'P_i + P_iA - P_iB_1P_{i-1} - P_{i-1}B_1P_i + P_{i-1}B_1P_{i-1} + N_1 = 0,$$
$$P_i(T) = N_3, \ i > 0, \ P_{-1} = 0.$$

By virtue of (2.1) a control of the j-th approximation v_j for $V_j = x'P_jx$ is given by the formula $v_j = -N_2^{-1}B'P_{j-1}x$ and in addition the differences $P - P_i$ and $I(u_0) - I(v_i)$ are values of $1/(i+1)!$ order.

2.2. Method of successive approximations for systems with small parameter. Assume that the relations (1.3) may be represented in the form

$$\inf_{u\in U} [L_uV+F_2(t,x,u)+\varepsilon\phi(t,x,V)]=0, \quad V(T,x)=F_1(x) \tag{2.2}$$

Let V_0 be a solution of the problem (2.2) with $\varepsilon=0$. Define approximations V_i, $i \geq 1$ in the following way

$$\inf_{u\in U} [L_uV_i+F_2(t,x,u)] = -\varepsilon\phi(t,x,V_{i-1}), \quad V_i(T,x) = F_1(x). \tag{2.3}$$

The i-th approximation control v_i is obtained as a solution of the problem (1.4) with V_i instead of V in the left side. Eqs.(1.8) are an example of the Eqs. (2.2) corresponding to the control system with small disturbances. In this case equations of successive approximations have the form

$$V_{it} + H(t,x,V_{it}) = -\frac{\varepsilon}{2}\operatorname{Tr}\sigma_1 V_{(i-1)xx}, \qquad V_i(T,x) = F_1(x). \tag{2.4}$$

Note that Eqs. (2.4) with the equation of zero approximation correspond to the deterministic problems of optimal control.

2.3. *Small control*. Let in the problem (1.1), (1.2) be

$$f = a(t,x) + \varepsilon q(t)u, \quad F_2 = \alpha(t,x).$$

Then Eq. (1.3) has a form (2.2) with

$$L_u V = L_0 V = V_t + a'V_x + \frac{1}{2}\operatorname{Tr}\sigma_1 V_{xx}, \quad \phi = \phi(t,V) = \inf_{u\in U}\left[V_x' qu\right].$$

Successive approximations V_i are determined by the relations

$$L_0V_0 + \alpha(t,x) = 0, \quad V_0(T,x) = F_1(x),$$
$$L_0V_i + \alpha(t,x) = -\varepsilon\phi(t,V_{i-1}), \quad V_i(t,x) = F_1(x), \quad i \geq 1. \tag{2.5}$$

In some cases Eqs.(2.5) may be solved analytically [8,10].On calculating V_i the quasi optimal synthesis of the i-th approximation $v_i(t,x)$ is derived from the condition

$$\inf_{u\in U}[V_{ix}' qu] = V_{ix}' qv_i. \tag{2.6}$$

Quality of the i-th approximation $v_i(t,x)$ is given by $I(v_i)$. Accuracy of the approximate synthesis (2.5), (2.6) for small ε is characterized by the value

$$I(v_i) - I(u_0) = O(\varepsilon^{i+1}).$$

In some cases procedure (2.5), (2.6) ensures the convergence of V_i under $i\to\infty$ to the exact solution V of Bellman equation (2.2) for arbitrary (finite) value ε and also the convergence $I(v_i)$ to $I(u_0)$.

3. Numerical and approximate methods of control synthesis in concrete problems. Numerical methods may be divided into two groups. The methods of the first group are founded as a rule on finite difference schemes for partial differential equation that are applied either directly to a Bellman equation or to relations approximating this equation. The relations (1.9), (1.10), (1.14), (1.19), (2.1), (2.3) are an example of such approximation. The special features of the Bellman equation numerical solution and some results for mechanical systems are given in

[13,23,24,36]. The methods of the second group consist of preliminary approximation of the initial continuous control problem by a problem with discrete time and finite numbers of states and control and consequent application of numerical procedures [13-15,28,30]. In [16] it's proposed to approximate the optimal cost function value by local solutions of a Bellman equation defined in some domain of a phase space. In addition if this domain is sufficiently large and the system starts from its depth then the probability of leaving the domain will be small. Below some results of numerical and approximate solutions of concrete optimal control problems are given.

3.1. Control of oscillations under stochastic disturbances. Consider the control that maximizes the probability of a system to be in a given region S_0 on the given time interval $[0,T]$. The applied finite difference scheme of the corresponding Bellman equation solution is effective also for another problems. The equations of a system have a form

$$\dot{x} = y, \ \dot{y} = -a^2x + bu + \sigma\xi \ , \ 0 \le t \le T. \tag{3.1}$$

Here a,b,σ,T are given and control u is a such that $|u| \le 1$. The initial condition for the system (3.1) is $(x,y) \in S_0$ for $t = 0$. Denote by $V(t,x,y)$ the Bellman function of this problem that equals to the maximum of probability of system (3.1) to be in S_0 on the interval $[t,T]$ under condition that $(x(t),y(t)) = (x,y) \in S_0$. Go over to new variables according to the formulae

$$t \to \sigma^2 t/(2b^2), \ x \to \sigma^4 x/(4b^3), \ y \to \sigma^2 y/2b, \ u \to bu, \ a^2 \to 4b^4a^2/\sigma^4, \ T \to \sigma^2 T/(2b^2).$$

Assume that in new variables domain S_0 is a square with the origin as a center and with the sides parallel to the coordinate axes. There are boundary condition $V(t,x,y) = 0$, $0 \le t \le T$ on the regular part of the boundary of S_0 and initial condition $V(T,x,y) = 1, (x,y) \in S_0$. Optimal control u_0 equals $u_0(t,x,y) = \mathrm{sign} V_y(t,x,y)$. Bellman equation has a form

$$V_t + yV_x - a^2xV_y + |V_y| + V_{yy} = 0.$$

Numerical solution of this equation was obtained in [17] by the fractional step method that leads to the following scheme. Let h_1, h_2, τ are approximation steps in x,y,t. Integer-valued indexes i,j,k vary within the limits $-N \le i, j \le N$, $K \ge k > 0$ and V_{ij}^k is a value of the function $V(t,x,y)$ for $x = ih_1$, $y = jh_2$, $t = k\tau$. The difference equations for V_{ij}^k following from Bellman equation have the form ($q = q(j) = 0$ for $j \ge 0$ and $q(j) = -1$ for $j < 0$)

$$\tau^{-1}(V_{ij}^k - V_{ij}^{k-0.5}) + jh_2h_1^{-1}(V_{i+1+q,j}^{k-0.5} - V_{i+q,j}^{k-0.5}) = 0,$$

$$\tau^{-1}(V_{ij}^{k-0.5} - V_{ij}^{k-1}) + h_2^{-2}(V_{i,j+1}^{k-1} - 2V_{i,j}^{k-1} + V_{i,j-1}^{k-1})$$

$$+(2h_2)^{-1}(u_{ij} - a^2 h_1 i)(V_{i,j+1}^{k-1} - V_{i,j-1}^{k-1}) = 0.$$

Here

$$u_{ij} = \mathrm{sign}(V_{i,j+1}^{k-0.5} - V_{i,j-1}^{k-0.5}).$$

Initial and boundary conditions lead to equalities

$$V_{-N,j_1}^{k} = V_{N,j_2}^{k} = V_{i_1,-N}^{k} = V_{i_1,N} = 0, \; K \ge k \ge 0$$

$$V_{ij}^{k} = 1, \; -N < i, \; j < N, \; -N \le i_1 \le N, \; -N \le j_1 < 0, \; 0 < j_2 \le N.$$

These equations for V_{ij}^k are solved numerically. Some results are given in Figure 1 where in the right are switching curves of the optimal control for $t = 0.6$, $a^2 = 0,1,2,3$ and level curves of Bellman function $V(t,x,y)$ for $a^2 = 1$, $t = 0.5$ (in the left). Besides $V(t,x,y) = V(t,-x,-y)$ and $u_0(t,x,y) = -u_0(t,-x,-y)$ that gives the possibility to define V and u_0 in another parts of the square S_0.

Figure 1:

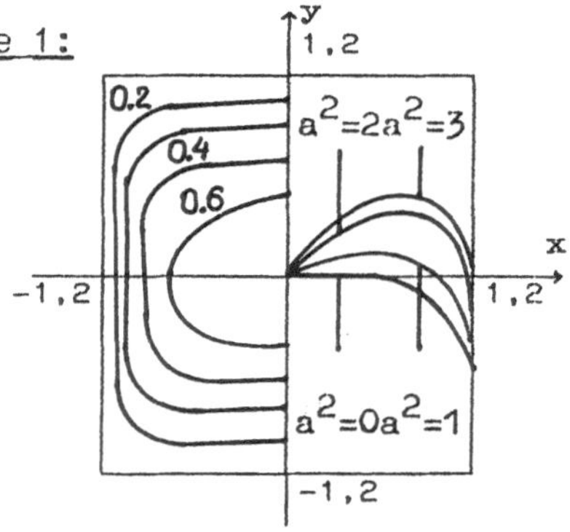

Figure 2:

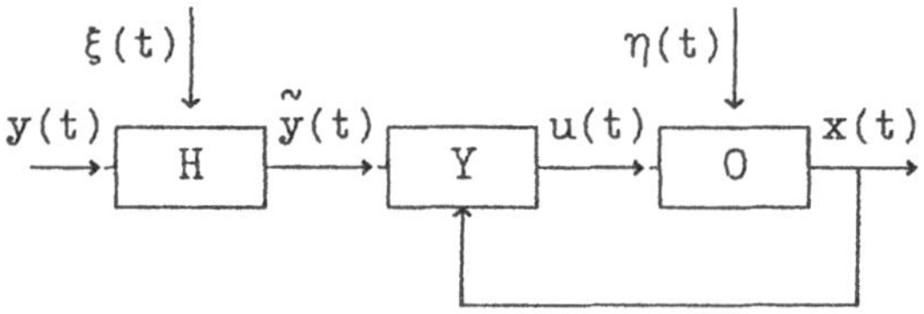

3.2. *Optimal tracking servo-system.* Let us consider the tracking servo-system shown in Figure 2. The input $y(t)$ is a symmetrical two-state Markov process ($y(t) = \pm 1$).Its a priori probabilities $p_t(\pm 1) = P(y(t) = \pm 1)$ satisfy Eqs.

$$\dot{p}_t(1) = -\dot{p}_t(-1) = -\mu p_t(1) + \mu p_t(-1)$$

The observed process $\bar{y}(t) = y(t) + \xi(t)$ is a mixture of an input process $y(t)$ and Gaussian white noise of intensity $\text{æ}/\varepsilon$. The control plant O is a servo-motor with constrained velocity disturbed by Gaussian white noise of intensity $\varepsilon\nu$. The behaviour of the plant O is described by a scalar equation

$$\dot{x} = u + \eta(t), \; |u| \le 1, \; M\eta(t) = 0, \; M\eta(t)\eta(t-\tau) = \varepsilon\nu\delta(\tau). \qquad (3.2)$$

The synthesis problem consists in finding a structure of the control device which on the basis of information about processes $\{\bar{y}(s), x(s): 0 \le s \le t\}$ for each moment t must form the control signal u(t) providing an optimal tracking of output coordinate x(t) after the input process y(t) in the sense of the minimum of functional

$$I(u) = M[\int_0^T (y(t) - x(t))^2 dt] \tag{3.3}$$

Let us denote by $w_t^+ = P(y(t) = 1|\bar{y}_0^t)$, $w_t^- = P(y(t) = -1|\bar{y}_0^t)$ the a posteriori probabilities of the states of the input process $y(t) = \pm 1$ in the presence of the observation $\bar{y}_0^t = \{\bar{y}(s): 0 \le s \le t\}$. According to [8] the difference $z(t) = w_t^+ - w_t^-$ satisfies the stochastic differential equation

$$\dot{z} = -2\mu z + \varepsilon(1 - z^2)\bar{y}(t)/æ. \tag{3.4}$$

The pair (x(t),z(t)) forms the vector of "phase coordinates" for the synthesis problem under consideration. From (3.2) - (3.4) it follows that the Bellman equation for this problem has the form

$$V_t + \min_{|u|\le 1}(uV_x) - 2\mu z V_z + \varepsilon\nu V_{xx}/2 + \varepsilon(1-z^2)^2 V_{zz}/2æ + x^2 - 2zx + 1 = 0,$$
$$0 \le t < T,\ -1<x,z<+1,\ V_x(t,\pm 1,z)=0,\ |V_z(t,x,\pm 1)|<\infty,\ V(T,x,z) = 0. \tag{3.5}$$

The diffusion terms of this equation contain a small parameter. Therefore for approximate solving (3.3) one can use the following scheme for calculating the successive approximations ($j = 0,1,\dots,V_{-1} = 0$)

$$V_{jt} - |V_{jx}| - 2\mu z V_{jz} + x^2 - 2zx + 1 = -\varepsilon(V_{j-1xx}/2 + (1 - z^2)^2 V_{j-1zz}/2æ)$$
$$v_j = -\operatorname{sign} V_{jx}.$$

By using corresponding calculations of the first two approximations for the stationary tracking problem (when the final time of tracking $T \to \infty$) the structural circuit of suboptimal tracking servo-system shown in

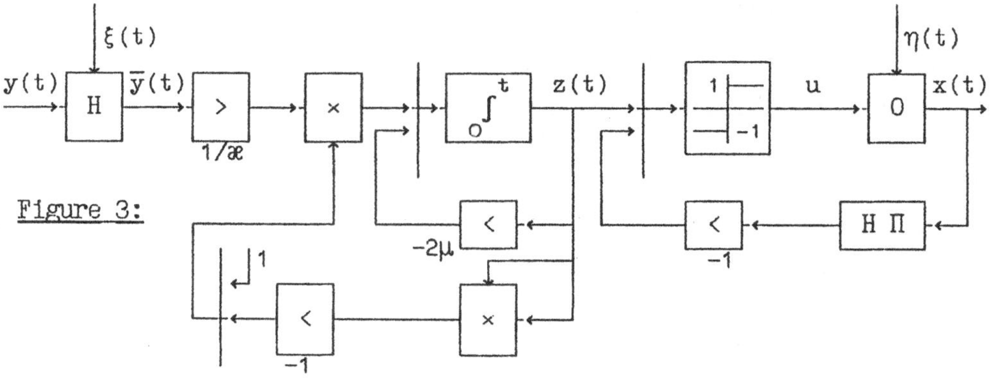

Figure 3:

Figure 3 was constructed in [8], where HΠ denotes the nonlinear converter

$$z = x + 2\varepsilon\mu x[(1 - x^2)^2 + \nu æ]/(1 - 4\mu^2 x^2)æ.$$

The quality of system shown in Figure 3 was estimated by numerically solving the Bellman equation (3.5) and the linear equation for performance index $Q(t,x,z)$ of the system in Figure 3

$$Q_t + v(t,x,z)Q_x - 2\mu z Q_z + \varepsilon\nu Q_{xx}/2 + \varepsilon(1 - z^2)^2 Q_{zz}/2æ +$$

$$+ x^2 - 2zx + 1 = 0, \; 0 \le t < T, \; -1 < x,z < +1, \; Q_x(t,\pm 1,z) = 0,$$

$$|Q_z(t,x,\mp 1)| < \infty, \; Q(T,x,z) = 0,$$

$$v(t,x,z) = \text{sign}\{x - z + 2\varepsilon\mu x[(1 - x^2)^2 + \nu æ]/(1 - 4\mu^2 x^2)æ\}.$$

Some results of the calculations carried out for $\varepsilon = \nu = 1$, $\mu = 0.45$, $æ = 5$, $t = T-4$ are shown in Figures 4,5 where continuous lines correspond to function V and dotted lines to Q. From the analysis of the curves in Figure 4,5 it follows that the circuit in Figure 3 insures the relative error of minimized functional

$$\delta I = [Q(t,x,z) - V(t,x,z)]/V(t,x,z) < 0.02.$$

Figure 4:

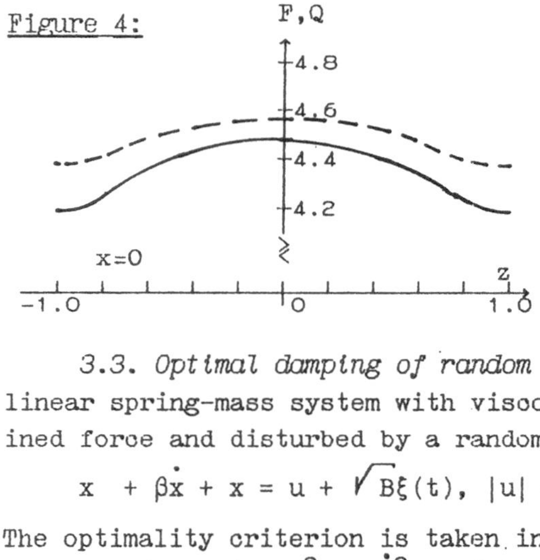

Figure 5:

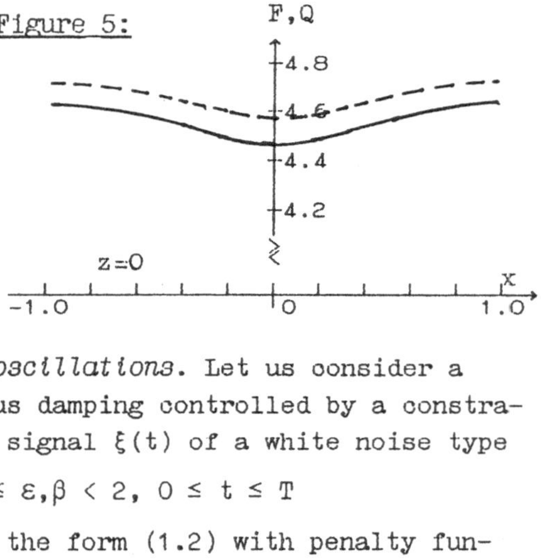

3.3. *Optimal damping of random oscillations.* Let us consider a linear spring-mass system with viscous damping controlled by a constrained force and disturbed by a random signal $\xi(t)$ of a white noise type

$$\ddot{x} + \beta\dot{x} + x = u + \sqrt{B}\xi(t), \; |u| \le \varepsilon, \beta < 2, \; 0 \le t \le T$$

The optimality criterion is taken in the form (1.2) with penalty functions $F_2(t,x,u) = x^2 + \alpha\dot{x}^2$, $F_1(x) = 0$. Using the dynamic programming algorithm (1.3), (1.4) we find optimal control

$$u_0(t,x,y) = -\varepsilon \,\text{sign} V_y(t,x,y), \tag{3.6}$$

where the Bellman function $V(t,x,y)$ $(y = \dot{x})$ satisfies the equation

$$\begin{gathered} V_t + yV_x - (x + \beta y)V_y + (B/2)V_{yy} = -x^2 - \alpha y^2 + \varepsilon|V_y|, \\ -\infty < x,y < +\infty, \; 0 \le t < T, \; V(T,x,y) = 0. \end{gathered} \tag{3.7}$$

This equation was solved numerically by using standard procedure of the net method [17]. The calculating scheme for Eq. (3.7) and some results of numerical solution of Eq. (3.7) were stated in detail in [11,37]. If a range of admissible control is small then the quantity ε in (3.7) is small parameter and for solving Eq. (3.7) it is possible to use successive approximations procedure (2.5). According to (2.5), (3.7) equations of zero and first approximation have the form

$$V_{0t} + yV_{0x} - (x + \beta y)V_{0y} + (B/2)V_{0yy} = -x^2 - \alpha y^2, \; V_0(T,x,y) = 0, \quad (3.8)$$

$$V_{1t} + yV_{1x} - (x + \beta y)V_{1y} + (B/2)V_{1yy} = -x^2 - \alpha y^2 + \varepsilon|V_{0y}|, \; V_1(T,x,y) = 0. \quad (3.9)$$

For constructing corresponding suboptimal control algorithms it is necessary to solve Eqs. (3.8), (3.9) and to substitute the obtained solutions V_0, V_1 into (3.6) instead of V. Eqs. (3.8),(3.9) and consequently the suboptimal control synthesis problem for zero and first approximations were solved in [8,10] in analytical form.

It is interesting to compare the optimal quantity $V(t,x,y)$ of considered criterion with the value $Q(t,x,y)$ of the same criterion when control $v\ (t,x,y)$ is used. The function $Q(t,x,y)$ is defined by relations

$$Q_t + yQ_x - (x + \beta y - v_0)Q_y + (B/2)Q_{yy} = -x^2 - \alpha y^2, \; Q(T,x,y) = 0 \quad (3.10)$$

The solutions of Eqs.(3.7),(3.10) were obtained numerically for $\alpha=\beta=B=1$, $\varepsilon=0.5$. The results are represented in Figure 6 where V is shown by the continuous line, Q by the dotted line and $\tau=T-t$ is stated.

Figure 6:

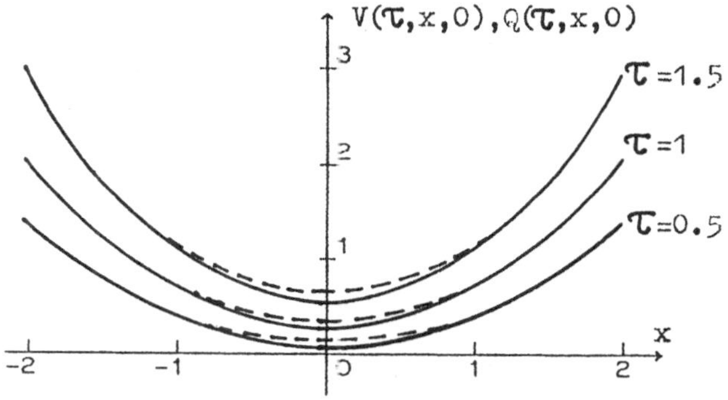

3.4. Optimal control by stochastic plant with unknown parameter. Let us consider a first-order aperiodic plant with an unknown inertia factor disturbed by Gaussian white noise ξ. Eq (1.1) in this case is a scalar one and has the form

$$\dot{x} = -\theta x + bu + \sqrt{\nu}\,\xi(t) \quad (3.11)$$

where θ is an unknown parameter, b,ν are given positive numbers. The control purpose consists in minimizing the functional (1.2) with penalty functions $F_1 = 0$, $F_2 = gx^2 + hu^2$, where g,h are given positive numbers. Let θ be a random value of Gaussian type with the mean θ_0 and variance εD_0. Let us write the optimal filtering equations for this case

$$dm(t) = -(D(t)x(t)/\nu)[dx(t) + (m(t)x(t) - bu(t))dt] \tag{3.12}$$

$$\dot{D}(t) = -D^2(t)x^2(t)/\nu, \quad m(0) = \theta_0, \quad D(0) = \varepsilon D_0 \tag{3.13}$$

The totality of scalar Eqs. (3.11)-(3.13) plays a part of a "dynamics" Eq.(1.1) for the synthesis problem under consideration. Let us do the substitution $D \to \varepsilon D$. Then using (3.11)-(3.13) one can write the Bellman Eq.(1.17) for this problem and obtain the optimal control in the form

$$u_0(t,x,m,D) = -bV_x(t,x,m,D)/2h \tag{3.14}$$

Here $V = V(t,x,m,D)$ is a solution of the following Cauchi problem

$$\begin{aligned} -V_t &= -mxV_x - b^2(V_x)^2/4h + gx^2 + \nu V_{xx}/2 \\ &-\varepsilon[DxV_{mx} + D^2X^2V_D/\nu - \varepsilon D^2x^2V_{mm}/2\nu], \quad V(T,x,m,D) = 0. \end{aligned} \tag{3.15}$$

If parameter ε has a small value then for solving Eq.(3.15) (and therefore according to (3.14) the synthesis problem) one can use small parameter method according to scheme (1.18)-(1.21). When $\varepsilon=0$ Eq. (3.15) has an exact solution

$$V_0(t,x,m) = f_0(t,m)x^2 + r_0(t,m) \tag{3.16}$$

$$f_0(t,m) = \frac{g[1-\exp(-2\gamma(T-t))]}{\gamma+m+(\gamma-m)\exp[-2\gamma(T-t)]}, \quad \gamma = (m^2+b^2g/h)^{1/2}$$

$$r_0(t,m) = \frac{g\nu(T-t)}{\gamma+m} - \frac{\nu h}{b^2}\ln\frac{2\gamma}{\gamma+m+(\gamma-m)\exp[-2\gamma(T-t)]}.$$

From (3.14), (3,16) follows the zero approximation control

$$v_0(t,x,m) = -bf_0(t,m)x/h. \tag{3.17}$$

The equation for the function V_1 from (1.18) has the form

$$-V_{1t} = -mxV_{1x} - b^2f_0(t,m)xV_{1x}/h + \nu V_{1xx}/2 - DxV_{0mx}, \quad V_1(T,x,m,D) = 0.$$

This equation has the following solution [8]

$$\begin{aligned} &V_1(t,x,m,D) = f_1(t,m,D)x^2 + r_1(t,m,D), \\ &f_1(t,m,D) = -2D\exp\{-2\int_0^T(m + b^2f_0(T-s,m)/h)ds\} \\ &\times\int_t^T f_{0m}(T-s,m)\exp\{2\int_s^T(m + bf_0^2(T-s_1,m)/h)ds_1\}ds, \end{aligned} \tag{3.18}$$

$$r_1(t,m,D)=\nu\int_t^T f_1(s,m,D)ds$$

From (3.14),(3.16),(3.18) it follows that first approximation control is given by the formula

$$v_1(t,x,m,D) = -bh^{-1}[f_0(t,m) + \varepsilon f_1(t,m,D)]x.$$

With the help of numerical methods a comparison was made between minimum value $V(t,x,m,D)$ of the considered performance index (obtained by solving Eq.(3.15)) and the values $Q_0(t,x,m,D)$ and $Q_1(t,x,m,D)$ of the same performance index for control v_0 and v_1 obtained by solving corresponding linear Eqs.

$$-Q_{it}= (-mx + v_i b)Q_{ix}+ hv_i^2+ gx^2+ \nu Q_{ixx} /2 - Dx(Q_{imx}+ DxQ_{iD}/\nu- DxQ_{imm}/2\nu),$$

$$Q\ (T,x,m,D) = 0,\ i = 0,1.$$

In Figure 7 the continuous lines correspond to V, dotted lines to Q_0 when problem parameters are $g=h=b=\nu=m=1$ and $\tau = T - t = 3$. In Figure 8 continuous lines correspond to V, dot-dash lines to Q_0, and dotted lines to Q_1 when $g = h = 1$, $b = 0.1$, $\nu = 5$, $\tau = 2.5$.

Figure 7:

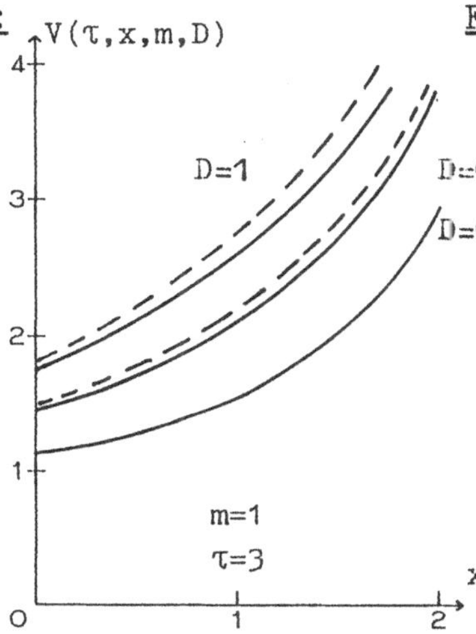

Figure 8:

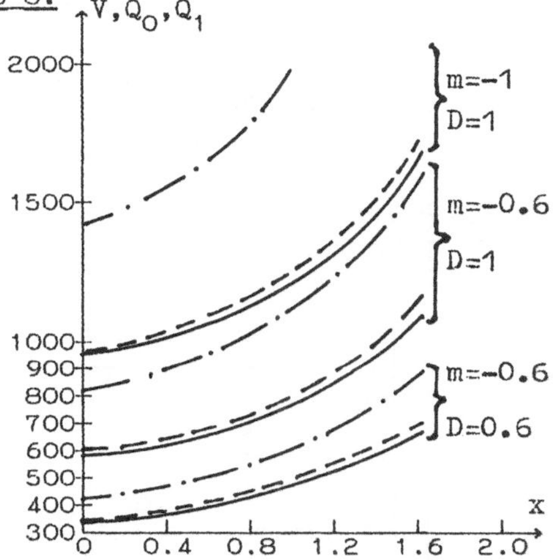

3.5. Optimal control by substance diffusion. In [8] the successive approximations scheme (2.5), (3.8), (3.9) was used for approximate synthesis of controls for some stochastic systems with distributed parameters. Let the object of control be a cylinder filled by a homogeneous porous medium, of length l and of base radius $r \ll l$ (that allows to ignore the radial changing of substance strength G and consider it depending only on $(t,x), 0 \le x \le l$). Suppose that at one side $(x = l)$ cylinder is closed and at the other side $(x = 0)$ the substance flow is set. Changing the flow one can influence substance strength $G(t,x)$. Suppose that along whole cylinder length random disturbance $\xi(t,x)$ influences the substance strength $G(t,x)$. Let $\xi(t,x)$ be a Gaussian white noise correlated in space variable. Then mathematical model of control plant has the form

$$G_t = a^2 G_{xx} + \xi(t,x),\ 0 < x < 1,\ a^2 = B/C, \tag{3.19}$$

$$G_x(t,0) = u,\ G_x(t,1) = 0, \tag{3.20}$$

$$M\xi(t,x) = 0,\ M\xi(t,x)\xi(\tau,y) = K(x,y)\delta(t-\tau). \tag{3.21}$$

Here B and C are diffusion and porous factors, $K(x,y)$ is given symmetrical positive defined function-kernel. For the object (3.19)-(3.21) it is necessary to synthesize control $u=u_0(t,G(t,x))$ minimizing the functional

$$I = M\left[\int_0^T\int_0^1\int_0^1 \theta(x,y)G(t,x)G(t,y)dxdydt\right] \tag{3.22}$$

when control action u is constrained by its absolute value $|u| \le \varepsilon$.

Using dynamic programming method to the problem (3.19)-(3.21) leads to the optimal control operator

$$u_0 = u_0(t,G(t,x)) = \varepsilon \operatorname{sign}\left[\left(\frac{\delta V}{\delta G(x,t)}\right)_{x=0}\right], \tag{3.23}$$

where loss functional $V(t,G(t,x))$ satisfies the Bellman equation with functional derivatives

$$-\frac{\partial V}{\partial t} = \int_0^1\int_0^1 \theta(x,y)G(t,x)G(t,y)dxdy + a^2\int_0^1 G(t,x)\frac{\partial^2}{\partial^2 x}\left(\frac{\delta V}{\delta G(t,x)}\right)dx$$

$$+ a^2\left[G(t,x)\frac{\partial}{\partial x}\left(\frac{\delta V}{\delta G(t,x)}\right)\right]_{x=0} - a^2\left[G(t,x)\frac{\partial}{\partial x}\left(\frac{\delta V}{\delta G(t,x)}\right)\right]_{x=1}$$

$$+ \frac{1}{2}\int_0^1\int_0^1 K(t,x,y)\frac{\delta^2 V}{\delta G(t,x)\,\delta G(t,y)}dxdy - \varepsilon a^2\left|\left(\frac{\delta V}{\delta G(t,x)}\right)_{x=0}\right|,$$

$$V(T,G(x,T)) = 0. \tag{3.24}$$

If the range of substance flow changing is small then ε in (3.24) is a small parameter and for approximate solving of (3.24) one can use successive approximations scheme (2.5), (3.8),(3.9). In this case as is established in [8] the functionals $V_0(t,G(t,x)), V_1(t,G(t,x))$, ... are calculated in the quadrature form. Suboptimal control operators $v_i(t,G(t,x))$ are obtained by means of substitution V_i in (3.23) instead of V. In particular control operator of zero approximation $v_0(t,G(t,x))$ has the form

$$v_0(t,G(t,x)) = \varepsilon \,\mathrm{sign}\left[\int_t^T d\tau \int_0^1\int_0^1\int_0^1 \theta(x,y)\Gamma(\tau,x,t,0)\Gamma(\tau,y,t,\bar{y})G(t,\bar{y})d\bar{y}dxdy\right],$$

where

$$\Gamma(\tau,x,t,y)= \frac{2}{l}\left[\frac{1}{2}+\sum_{n=1}^{\infty}\exp\left[-\left(\frac{\pi n}{l}\right)^2 a^2(\tau-t)\right]\cos\frac{\pi n}{l}x\cos\frac{\pi n}{l}y\right]$$

is the Green function for boundary problem (3.19), (3.20).

Conclusion. Approximate synthesis methods make it possible to obtain control algorithms in analytical form that is important for practical using of stochastic optimal control methods in concrete problems. Besides the results of numerical analysis carried out for concrete optimal control problems show a high efficiency of the small parameter and successful approximations methods in the cases when parameter ε has the same value order as another parameters of the synthesis problem. Moreover the calculation results represented in Figures 6,7 show that it is possible when even zero approximation control insures a control quality close to the optimal one. In other cases a zero approximation control is insufficient and for high quality control it is necessary to use next (higher-order) approximations which is illustrated by results represented in Figures 4, 5, 8

References.

1. Bellman R. Dynamic programming. Princeton, 1957.

2. Glizer V.Ya. Asymptotics of solution of one singular perturbated Cauchy problem in theory of optimal linear filtering // Izvestia vuzov (matematica), 1984, 12, 55-58.

3. Kolmanovskii V.B. Suboptimal programming control for some stochastic systems // Differential equations, 1980, 16, 5.

4. Kolmanovskii V.B. On stabilizing of some nonlinear systems // Appl. Math.& Mech., 1987, 51, 3, 395-402.

5. Kolmanovskii V.B. Some control problems for the systems with a small parameter // Supplement to the book: Donchev A. Optimal control systems. Moscow: Mir, 1987.

6. Kolmanovskii V.B. and L.Ye. Shaykhet. Optimal control synthesis for some mechanical systems with random disturbances // Eng. Cybern., 1983,2.

7. Kolmanovskii V.B. and L.Ye. Shaykhet. Optimal control for stochastic hereditary systems with uncertain parameters // Kubern., 1988,2.

8. Kolosov G.Ye. Optimal design of automatic systems under stochastic disturbances. Nayka, Moscow, 1984.

9. Kolosov G.Ye.Some asymptotical and numerical synthesis methods for stochastic optimal control systems // Technische Hochschule Leipzig Wissenschaftliche Zeitschrift, 1988, 3, 161-170.

10. Kolosov G.Ye., and R.L. Stratonovich. On one asymptotic method of solution of problems of optimal controller synthesis // Autom. & Remote Control, 1964, 25, 12.

11. Kolosov G.Ye., and M.M. Sharov. On a numerical method to develop stochastic systems of optimal control // Autom. & Remote Control. 1987, 8, 98-105.

12. Krasovskii A.A. Phase space and statistical theory of dynamical systems. Nauka, Moscow, 1974.
13. Kushner H.J. Probability methods for approximations in stochastic control and for elliptic equations. N.Y. etc, 1977.
14. Lebedev A.A., Krasilshikov M.N., and V.V. Malishev. Optimal control of motion of space flying machines. Mashinostr., Moscow, 1974.
15. Lebedev A.A., Bobronnikov V.T., Krasilshikov M.N., and V.V. Malishev. Statistical dynamics and optimization of control of flying machines. Mashinostr., Moscow, 1985.
16. Ovseyevich A.I. Local Bellman principle in optimal control problems // Eng. Cybern., 1981, 4, 3-9.
17. Samarokii A.A. Introduction in difference-scheme theory. Nauka, 1971.
18. Sobolev V.A. Integral manifolds, singular disturbance and optimal control // Ukr. Matem. J., 1987, 39, 1, 111-116.
19. Stratonovich R.L. Conditional Markov Processes and Their Applications to the Theory of Optimal Control.Am.Els.Co.,Inc., N.Y., 1968.
20. Fleming W.H. Stochastic control for small noise intensities. SIAM J. Control, 1971, 9, 3, 437-515.
21. Khasminskii R.Z. Stability of systems of differential equations with random disturbances of its parameters. Nauka, Moscow, 1969.
22. Chernons'ko F.L. Optimization problems for mechanical systems // Uspekhi mekh., 1979, 2, 1, 3-36.
23. Chernous'ko F.L., and V.B. Kolmanovskii. Computational and approximate methods of optimal control // Math. anal., VINITI AN SSSR, 1977, 14, Total of sci. and eng.
24. Chernous'ko F.L., and V.B. Kolmanovskii. Optimal control under stochastic disturbances. Nauka, Moscow, 1978.
25. Shaykhet L. Ye. On approximate synthesis of quadratic and nonquadratic control problems // Theory of stochastic processes, 1981, 9.
26. Bell D.J., Jacobson D.H. Singular optimal control problems. New York. Academic Press, 1975.
27. Bensoussan A. Perturbation methods in optimal control. New York. Dunod and Wiley, 1986.
28. Bensoussan A., Runggaldier W. An approximation method for stochastic control problems with partial observation of the state-a method for constructing ε-optimal controls // Acta Appl. Math.1987,10,145-170.
29. Chun-Ping Tsai. Perturbed stochastic linear regulator problems. // SIAM J. Control and Optimization, 1978, 16, 3, 396-410.
30. Di Masi, Pratelli G.B., Runggaldier W.J. An approximation for the non linear filtering problem with error bound // Stochastics, 1985, 14, 247-271.
31. Fleming W.H. Some Markovian optimization problems. J.Math. and Mech., 1963, 12, 1.
32. Holland C. Gaussian open loop control problems // SIAM J.Control, 1975, 13.
33. Khalil H.K., Cajic Z. Near-optimal regulators for stochastic linear singularly perturbed systems // IEEE Trans. Autom. Control, 1984, 29, 531-541.
34. Kokotovich P.V. Applications of singular perturbations techniques to control problems // SIAM review, 1984, 26, 4, 501-550.
35. Kokotovich P.V., Khalil H.K., O'Reilly Singular perturbation methods in control: analysis and design. London: Academic Press, 1986.
36. Kolmanovskii V.B. Control and estimation in stochastic hereditary systems // Preprints of 2 IFAC Symp. on stoch.control. Vilnius, 1986, 227-231.
37. Kolosov G.Ye. Numerical synthesis of optimal control for some stochastic systems // Preprints of 2 IFAC Symp. on stoch.control. Vilnius, 1986, 301-306.
38. Singh P.R. The linear-quadratic Gaussian problem for singularly perturbed systems // Int. J. Systems Sci. 1982, 13, 1, 93-100.

Convergence Rates for Regularized Nonlinear Illposed Problems*

K. KUNISCH AND G. GEYMAYER
INSTITUT FÜR MATHEMATIK, TECHNISCHE UNIVERSITÄT GRAZ
A-8010 GRAZ, AUSTRIA

Abstract. Convergence and rate of convergence are studied for nonlinear illposed inverse problems that are stabilized by means of Tikhonov regularization while the parameter space as well as the parameter-to-output mapping are discretized. The theoretical results are illustrated by means of numerical examples.

1. INTRODUCTION

In this contribution we focus on nonlinear illposed inverse problems of the type

$$F(x) = y_0, \tag{1.1}$$

where $F : D(F) \subset X \to Y$ is a nonlinear operator between Hilbert spaces X and Y. The problem consists in inverting F at y_0 in a stable manner without making assumptions on the continuous invertibility of F at y_0 and while allowing errors in the "data" y_0. To address this problem the regularized least squares formulation

$$\min |F(x) - y_\delta|_Y^2 + \alpha |x - x^*|_X^2 \quad \text{over} \quad D(F) \tag{1.2}$$

is used. Here y_δ denotes the noisy data which are assumed to satisfy an a–priori estimate of the type

$$|y_0 - y_\delta| \leq \delta, \tag{1.3}$$

and x^* stands for an estimate to a solution of (1.1). The problem of an adequate choice of α in terms of δ such that the solutions x_α^δ of (1.2) converge and also converge with a certain rate as $\delta \to 0$ was extensively studied for the case when F is linear (see e.g. [G,M] and the references given there) and has recently been investigated for nonlinear F for instance in [EKN,N]. If the domain and the range of F are infinite dimensional, then the optimization problems (1.2) are infinite dimensional as well, and any numerical approach to solve (1.2) will require a discretization of $D(F)$ as well as of the mapping F. In this note we address the problem of convergence and of rate of convergence of the solutions of the fully discretized version of (1.2) as $\delta, \alpha \to 0$ and as the discretization indices tend to infinity. The case when only $D(F)$ is discretized was already treated in [N]. Moreover we give selected numerical results for parameter estimation in a two point boundary value problem. Such problems are wellknown to be nonlinear illposed inverse problems. We shall illustrate that the theoretical results on the rate of convergence can be observed numerically and we shall illustrate the necessity of the hypotheses that are made. Many additional numerical results can be found in [Ge,GK]. The techniques that are required for the proof of the rate of convergence are strongly related to those in [EKN,N].

---

*Supported in part by the Fonds zur Förderung der wissenschaftlichen Forschung, Austria, under S3206.

2. Convergence and Rate of Convergence

The following specifications will be necessary:

X, Y are Hilbert spaces,
$F : D(F) \subset X \to Y$, with $D(F)$ closed and convex,
F is continuous and weakly sequentially closed, i.e. for any sequence $\{x_n\}$ in $D(F)$, $x_n \rightharpoonup x$ and $F(x_n) \rightharpoonup y$ imply $x \in D(F)$ and $F(x) = y$,
$y_0 \in R(F)$,
$\{X_n\}_{n=1}^{\infty}$ is a squence of finite dimensional subspaces of X,
$P_n : X \to X_n$ are the orthogonal projections, which satisfy $P_n x \to x$ for all $x \in X$,
$C_n := D(F) \cap X_n$, $P_{C_n} : X \to C_n$ is the metric projection,
$F^N : D(F) \to Y$ are continuous operators for $N = 1, 2, \ldots$.

Here and below we denote by '$\to$' strong and by '$\rightharpoonup$' weak convergence in a Hilbert space and $D(F)$ and $R(F)$ stand for the domain and the range of F. Due to the assumption that $y_0 \in R(F)$, the existence of a solution to $F(x) = y_0$ is trivially satisfied. Henceforth the focus will be on solutions to (1.1) which are closest to the estimator x^*. We define x_0 to be an x^*-minimum norm solution (x^*-MNS) of (1.1) if

$$F(x_0) = y_0$$

and

$$|x_0 - x^*| = \min\{|x - x^*| : F(x) = y_0\}.$$

The existence of an x^*-MNS is a consequence of the weak sequential closedness assumption. The motivation for the concept of x^*-MNS to (1.1) will follow from Proposition 1 below.

To solve (1.1) numerically with possibly noisy data y_δ we introduce the regularized finite dimensional problems

$$(\mathcal{P}) \qquad \min |F^N(x) - y_\delta|^2 + \alpha |x - P_n x^*|^2 \quad \text{over} \quad x \in C_n = D(F) \cap X_n;$$

where we tacitely assume that the range of F^N is finite dimensional, although this is not further necessary within this section. In $(\mathcal{P})$ we did not introduce additional notation for discretization of y_δ. Rather the data y_δ can be considered to be elements of finite dimensional spaces, converging to y_0 as $\delta \to 0$. Due to the assumptions on F^N and $D(F)$, it is simple to see that $(\mathcal{P})$ has a solution for any $\alpha > 0$, N, n and x^*. In our analysis we shall not insist on exact solutions to $(\mathcal{P})$ but rather we analyze the convergence of elements $x_{\alpha,n,N}^{\delta,\eta}$ which satisfy

$$(2.1) \quad |F^N(x_{\alpha,n,N}^{\delta,\eta}) - y_\delta|^2 + \alpha |x_{\alpha,n,N}^{\delta,\eta} - P_n x^*|^2 \le |F^N(x) - y_\delta|^2 + \alpha |x - P_n x^*|^2 + \eta$$

for all $x \in C_n$, where $\eta > 0$. Asymptotic expressions involving $\alpha, \delta, \eta, n, N$ will always be understood in the sense that $\alpha, \delta, \eta \to 0$ and $n, N \to \infty$. It will be shown next that weak cluster points of $x_{\alpha,n,N}^{\delta,\eta}$ are strong cluster points and that they are x^*-MNS.

PROPOSITION 1. *Let x_0 be an x^*-MNS of (1.1) and let $\alpha = \alpha(n, N, \delta, \eta)$ be such that $\alpha \to 0$, $\eta/\alpha \to 0$, $\delta^2/\alpha \to 0$ and $|F^N(P_{C_n}x_0) - y_0|^2/\alpha \to 0$ as $\delta, \eta \to 0$ and $N, n \to \infty$. Assume further that $F^N \to F$ uniformly on bounded subsets of $D(F)$ and that $P_{C_n}x_0 \to x_0$. Then every sequence $\{x^{\delta_k,\eta_k}_{\alpha_k,n_k,N_k}\}$ (with $\delta_k, \eta_k \to 0$; $N_k, n_k \to \infty$ for $k \to \infty$, $\alpha_k := \alpha(n_k, N_k, \delta_k, \eta_k)$ and $x^{\delta_k,\eta_k}_{\alpha_k,n_k,N_k}$ satisfying (2.1)), has a strongly convergent subsequence and the limit of every convergent subsequence of $x^{\delta,\eta}_{\alpha,n,N}$ is an x^*-MNS of $F(x) = y_0$.*

The proof can be given with standard arguments. We refer to [N] and the discussion given there. Concerning the hypotheses of the proposition we refer to [Mo, Lemma 1.2 and 1.5] where conditions on X_n are given that imply $\lim_{n\to\infty} P_{C_n}x_0 = x_0$. With respect to existence of values for α satisfying $|F^N(P_{C_n}x_0) - y_0|^2/\alpha \to 0$, one observes that

$$\frac{1}{\alpha}|F^N(P_{C_n}x_0) - y_0|^2 \leq \frac{2}{\alpha}|F^N(P_{C_n}x_0) - F(P_{C_n}x_0)|^2 + \frac{2}{\alpha}|F(P_{C_n}x_0) - y_0|^2.$$

For appropriate choice of α the middle term converges to zero, if $\lim_{N\to\infty}|F^N(x) - F(x)| = 0$ uniformly on bounded subsets of X and the last term converges to zero due to continuity of F and $\lim_{n\to\infty} P_{C_n}x_0 = x_0$.

We proceed with a result on the rate of convergence of $x^{\delta,\eta}_{\alpha,n,N}$ as $\delta, \alpha, \eta \to 0$, and $N, n \to \infty$. Some hypotheses are summarized first. They involve an x^*-MNS x_0.

- (H1) F is Fréchet differentiable,
- (H2) there exists $L > 0$ such that $\|F'(x_0) - F'(x)\| \leq L|x_0 - x|$, for all $x \in B_p(x_0) \cap D(F)$, where $p > 2|x_0 - x^*|$,
- (H3) there exists $\omega \in Y$ such that $x_0 - x^* = F'(x_0)^*\omega$,
- (H4) $L|\omega| < 1$,
- (H5) $|x_0 - P_n x_0| = O(\tilde{\gamma}_n)$ with $\lim \tilde{\gamma}_n = 0$,
- (H6) $|F(x) - F^N(x)| = 0(\varepsilon_N)$ for all $x \in B_p(x_0) \cap D(F)$, $p > 2|x_0 - x^*|$, with $\lim_{N\to\infty} \varepsilon_N = 0$.

Here $B_p(x_0) = \{x : |x - x_0| < p\}$ and $F'(x_0)^*$ stands for the adjoint of the Fréchet derivative of F at x_0. Furthermore we put

$$\gamma_n := \|F'(x_0)(I - P_n)\|,$$

and observe that $\lim_{n\to\infty} \gamma_n = 0$ if $F'(x_0)$ is a compact operator. In the following theorem it will be implicitly assumed that the interior of $D(F)$, int $D(F)$, is nonempty.

THEOREM 2. *Assume that x_0 is an x^*-MNS of $F(x) = y_0$, that (H1) - (H6) are satisfied with $x_0 \in$ int $D(F)$ and that $F'(x_0)$ is compact. If in addition $\eta = O(\delta^2 + \gamma_n^2\tilde{\gamma}_n^2)$ and $\alpha \sim \max(\delta, \tilde{\gamma}_n^2, \tilde{\gamma}_n, \gamma_n, \varepsilon_N)$, then*

$$|x^{\delta,\eta}_{\alpha,n,N} - x_0| = O(\sqrt{\delta} + \tilde{\gamma}_n + \sqrt{\tilde{\gamma}_n\gamma_n} + \sqrt{\varepsilon_N}).$$

PROOF: The first objective is to derive estimates on $|F(P_n x_0)-y_\delta|$, $|F^N(P_N x_0)-y_\delta|$ and to show that $z := x_{\alpha,n,N}^{\delta,\eta} \in B_p(x_0)$ for all n, N sufficiently large and α, δ, η sufficiently small. Since $x_0 \in \text{int } D(F)$, (H5) implies that

$$P_n x_0 \in C_n \tag{2.2}$$

for all n sufficiently large. Let us define

$$r_n := F(P_n x_0) - F(x_0) - F'(x_0)(P_n x_0 - x_0),$$

where here and below it is assumed that n is sufficiently large so that (2.2) holds and that $|P_n x_0 - x_0| < p$ with p as in (H2). Then by (H2) one finds

$$|r_n| \leq \frac{L}{2}|x_0 - P_n x_0|^2. \tag{2.3}$$

By (2.3) and due to $F(x_0) = y_0$ it follows that

$$\begin{aligned} |F(P_n x_0) - y_\delta| &= |r_n + F'(x_0)(P_n - I)x_0 + F(x_0) - y_\delta| \\ &\leq \frac{L}{2}|x_0 - P_n x_0|^2 + \gamma_n |x_0 - P_n x_0| + \delta \\ &= O(\tilde{\gamma}_n^2 + \gamma_n \tilde{\gamma}_n + \delta), \end{aligned} \tag{2.4}$$

where for the last estimate we used (H5). Moreover, by (H5), (H6) and (2.4)

$$|F^N(P_n x_0) - y_\delta \leq O(\tilde{\gamma}_n^2 + \gamma_n \tilde{\gamma}_n + \delta + \varepsilon_N) \tag{2.5}$$

holds for all n sufficiently large. Due to (2.1), (2.2), (2.5) and the assumption on η one obtains

$$\begin{aligned} |F^N(z) - y_\delta|^2 + \alpha|z - P_n x^*|^2 &\leq |F^N(P_n x_0) - y_\delta|^2 + \alpha|P_n x_0 - P_n x^*|^2 + \eta \\ &\leq O(\tilde{\gamma}_n^4 + \gamma_n^2 \tilde{\gamma}_n^2 + \delta^2 + \varepsilon_N^2) + \alpha|x_0 - x^*|^2, \end{aligned} \tag{2.6}$$

and hence

$$|z - x_0| \leq |z - P_n x^*| + |P_n x^* - x_0| \leq \frac{1}{\sqrt{\alpha}} O(\tilde{\gamma}_n^2 + \gamma_n \tilde{\gamma}_n + \delta + \varepsilon_N) + |P_n x^* - x^*| + 2|x_0 - x^*|.$$

Since $\alpha \sim \max(\delta, \tilde{\gamma}_n^2, \gamma_n \tilde{\gamma}_n, \varepsilon_N)$ it follows that $z \in B_p(x_0)$ for all n, N sufficiently large and δ sufficiently small.

These estimates conclude the verification of the claim at the beginning of the proof. Using the fact that $|a|^2 - |b|^2 = 2\langle a, a - b\rangle - |a - b|^2$ for all $a, b \in X$ one deduces from (2.6) that

$$\begin{aligned} |F^N(z) - y_\delta|^2 + \alpha|P_n x_0 - z|^2 \leq\ & O(\tilde{\gamma}_n^4 + \tilde{\gamma}_n^2 \gamma_n^2 + \delta^2 + \varepsilon_N^2) \\ & + 2\alpha\langle P_n x_0 - P_n x^*, P_n x_0 - z\rangle, \end{aligned} \tag{2.7}$$

which further implies the estimate

$$
(2.8)\qquad \begin{aligned}|F^N(z)-y_\delta|^2+\alpha|z-x_0|^2 &\le O(\tilde\gamma_n^4+\tilde\gamma_n^2\gamma_n^2+\delta^2+\varepsilon_N^2)+\alpha|P_nx_0-x_0|^2\\ &\quad+2\alpha\langle x_0-x^*,P_nx_0-z\rangle\\ &=O(\rho)+2\alpha\langle x_0-x^*,P_nx_0-z\rangle,\end{aligned}
$$

with $\rho=\tilde\gamma_n^4+\gamma_n^2\tilde\gamma_n^2+\delta^2+\varepsilon_N^2+\alpha\tilde\gamma_n^2$. For the next estimate one employs (H3):

$$
(2.9)\qquad \begin{aligned}|F^N(z)-y_\delta|^2+\alpha|z-x_0|^2 &\le O(\rho)+2\alpha\langle\omega,F'(x_0)^*(P_nx_0-x_0+x_0-z)\rangle\\ &=O(\rho)+2\alpha|\omega|\gamma_n\tilde\gamma_n+2\alpha\langle\omega,F'(x_0)^*(x_0-z)\rangle.\end{aligned}
$$

Due to (H2) and the fact that $z\in B_p(x_0)$ for all n,N sufficiently large, one can use Taylor's theorem to obtain ρ_z satisfying

$$
F(z)=F(x_0)+F'(x_0)(z-x_0)+\rho_z
$$

with

$$
|\rho_z|\le\frac{1}{2}L|z-x_0|^2,
$$

and hence from (2.9)

$$
(2.10)\qquad \begin{aligned}&|F^N(z)-y_\delta|^2+\alpha|z-x_0|^2\le O(\rho)+2\alpha|\omega|\gamma_n\tilde\gamma_n+2\alpha\langle\omega,F(x_0)-F(z)+\rho_z\rangle\\ &=O(\rho)+2\alpha|\omega|\gamma_n\tilde\gamma_n+2\alpha|\omega|\delta+2\alpha|\omega|\,|y_\delta-F(z)|+\alpha L|\omega|\,|z-x_0|^2,\end{aligned}
$$

from which it follows that

$$
(2.11)\qquad \begin{aligned}&|F^N(z)-y_\delta|^2+\alpha(1-L|\omega|)|z-x_0|^2\\ &\qquad\le O(\rho+\alpha\gamma_n\tilde\gamma_n+\alpha\delta)+2\alpha|\omega|\,|y_\delta-F^N(z)|+2\alpha|\omega|\varepsilon_N\\ &\qquad=O(\rho+\alpha\gamma_n\tilde\gamma_n+\alpha\delta+\alpha\varepsilon_N)+2\alpha|\omega|\,|F^N(z)-y_\delta|.\end{aligned}
$$

From (H4) and (2.11) with the second term on the left hand side eliminated one obtains

$$
(2.12)\qquad |F^N(z)-y_\delta|\le O(\sqrt{\rho}+\sqrt{\alpha\gamma_n\tilde\gamma_n}+\sqrt{\alpha\delta}+\sqrt{\alpha\varepsilon_N})+2\alpha|\omega|,
$$

and (2.11) together with (2.12) and (H4) imply

$$
\begin{aligned}|z-x_0|^2&\le O(\frac{\rho}{\alpha}+\gamma_n\tilde\gamma_n+\delta+\varepsilon_n)+O(\sqrt{\rho}+\sqrt{\alpha\gamma_n\tilde\gamma_n}+\sqrt{\alpha\delta}+\sqrt{\alpha\varepsilon_N}+\alpha)\\ &=O(\tilde\gamma_n^2+\gamma_n\tilde\gamma_n+\delta+\varepsilon_N).\end{aligned}
$$

This gives the desired estimate.

REMARKS:

(i) In applications, $\tilde\gamma_n$ in (H5) will converge to zero with a certain rate if a-priori smoothness properties of x_0 are known [KW]. (H5) can be replaced by a requirement on the estimator x^*. In fact, assume that

$$
\text{(H5')}\qquad |x^*-P_nx^*|=O(\tilde\gamma_n)
$$

and that (H3) holds. Then the following estimate can be used to replace (H5) [N]:

$$|x_0 - P_n x_0| = |x^* - P_n x^* + (I - P_n)F'(x_0)^*\omega| \le O(\tilde{\gamma}_n) + O(\gamma_n).$$

(ii) If it is known from other considerations as for instance from Proposition 1 that $x_{\alpha,n,N}^{\delta,\eta} \to x_0$, then (H2) and (H6) need only to hold in some neighborhood of x_0.

(iii) As a consequence of Theorem 2 it follows that if x_0 is not unique as an x^*-MNS of (1.1), that there can only exist one such x^*-MNS which satisfies the hypotheses of this theorem.

(iv) If F is twice continuously differentiable then (H2) and (H4) can be replaced by

$$\text{(H4')} \qquad 2\langle \omega, \int_0^1 (1-t)F''[x_0 + t(z - x_0)](z - x_0)^2 dt \rangle \le \tilde{\rho}|z - x_0|^2, \text{ with } \tilde{\rho} < 1.$$

The hypotheses (H2) and (H4) where used twice in the proof of Theorem 2. To estimate r_n one uses a version of the mean value theorem to establish the existence of $\tau_n \in [0,1]$ such that

$$r_n \le \frac{1}{2}|F''(x_0 + \tau_n(P_n x_0 - x_0))(P_n x_0 - x_0)^2|.$$

Since $\lim_{n\to\infty} P_n x_0 = x_0$ and since F'' is bounded in a neighborhood of x_0 there exists $\tilde{L}$ such that $r_n \le \tilde{L}|x_0 - P_n x_0|^2$. Secondly, due to (H4') the term $2\langle \omega, \rho_z \rangle$ in (2.10) can be bounded by $\tilde{\rho}|z - x_0|^2$ and in the remaining estimates $L|\omega| < 1$ is replaced by $\tilde{\rho} < 1$.

3. Numerical Examples

We shall illustrate Theorem 2 by considering the illposed problem of estimating the diffusion coefficient $a \in H^1(0,1)$, $a \ge \nu > 0$ in

$$\text{(3.1)} \qquad \begin{aligned} -(au_x)_x + cu &= f \quad \text{in} \quad (0,1), \\ u(0) = u(1) &= 0 \end{aligned}$$

from noisy data $y_\delta \in L^2(0,1)$ here it is assumed that $f \in L^2(0,1)$ and $c \in L^2(0,1)$, $c \ge 0$ a.e. It is assumed that the unperturbed measurement y_0 is attainable, i.e. that there exists $a \in H^1(0,1)$, $a \ge \nu$ such that $u(a) = y_0$. To realize this parameter estimation problem as a special case of the general theory one defines the operator F by $F : D(F) = \{a \in H^1(0,1) : a \ge \nu\} \to L^2(0,1)$, with

$$F(a) = u(a),$$

where $u(a)$ denotes the solution of (3.1) and puts $X = H^1(0,1)$ and $Y = L^2(0,1)$. It is not hard to argue that F is continuous, weakly sequentially closed and twice continuously differentiable. (See [EKN,N] for details and references). Henceforth we fix an estimator a^* and an a^*-MNS a_0 with $a_0 > \nu$ so that $a_0 \in \text{int } D(F)$. To discuss (H3) one introduces the Neumann operator $B : D(B) = \{\varphi \in H^2(0,1) : \varphi'(0) = \varphi'(1) = 0\} \rightarrow L^2(0,1)$ defined by $B\varphi = -\varphi_{xx} + \varphi$. Condition (H3) takes the form

$$a_0 - a^* = -B^{-1}(u_x(a_0)(A(a_0)^{-1}\omega)_x),$$

for some $\omega \in L^2(0,1)$, where for $a \in D(F)$, $A(a) : H_0^1 \cap H^2 \rightarrow L^2$ is given by $A(a)u = -(au_x)_x + cu$. Such an element ω exists if $a_0 - a^* \in H^3 \cap D(B)$ and $B(a^* - a_0) = u_x(a_0)(A(a_0)^{-1}\omega)_x$ or equivalently

$$\bar{\omega} := \int_0^{\cdot} \frac{B(a^* - a_0)}{u_x(a_0)}(s)ds \in H^2 \cap H_0^1, \tag{3.2}$$

and it is given by

$$\omega = A(a_0)\bar{\omega}. \tag{3.3}$$

As described at the end of the previous section (H2), (H4) can be replaced by (H4'). In [EKN] it was shown that (H4') is satisfied provided that

$$\frac{25}{2}\|A(a_0)^{-1}\|_{\mathcal{L}(L^2,H^2\cap H_0^1)}|\bar{\omega}|_{L^2}|u(a_0)|_{H^2} < 1. \tag{3.4}$$

Finally it is simple to check that $F'(a_0)$ is compact. Thus all hypotheses that do not involve discretization are satisfied if (3.2) - (3.4) hold. To turn to the discretization of this infinite dimensional nonlinear inverse problem let $S_n = \{\frac{i}{n}\}_{i=0}^n$ be a sequence of uniform grids on $[0,1]$, with $n = 1,2,\ldots$. We shall discretize the coefficient- and the statespace of (3.1) over the same grid and we shall use $n = N$. Let $X_n \subset H^1$ and $Y_n \subset H_0^1 \cap H^2$ denote the canonical spaces of linear- respectively cubic B-splines with respect to the grid S_n [BK, Appendix], where the cubic B-splines are modified so as to satisfy the boundary conditions. For $a \in D(F)$ let $F^n(a) = u^n(a)$ denote the Galerkin approximation to (3.1), i.e.

$$\langle au_x^n, v_x^n\rangle + \langle cu^n(a), v^n\rangle = \langle f, v^n\rangle \quad \text{for all} \quad v^n \in Y_n,$$

where $\langle\cdot,\cdot\rangle$ denotes the inner product in L^2. Finally let $P_n : H^1(0,1) \rightarrow X_n$ denote the orthogonal projections. From wellknown approximation properties of X_n and Y_n it follows that $\lim_{n\rightarrow\infty} P_n a = a$ for all $a \in H^1(0,1)$ and that $\gamma_n = O(\frac{1}{n})$. It is simple to check that $a \rightarrow F^N(a) = u^N(a)$ is continuous for $a \in D(F)$. Moreover for $a_0 \in H^2(0,1)$ or $a^* \in H^2(0,1)$ as will be assumed henceforth we have $\tilde{\gamma}_n = O(\frac{1}{n})$. With standard Galerkin techniques one can show that $\varepsilon_n = O(\frac{1}{n^2})$. Hence (H5), respectively (H5') and (H6) are satisfied. Thus the best rate that we can obtain with our estimates is $O(\frac{1}{n})$, if $\eta = O(\frac{1}{n^4})$ and $\alpha \sim \delta = O(\frac{1}{n^2})$. For the numerical results to be presented below X_n and Y_n are chosen as explained above. If Y_n was replaced by linear splines on the same grid then the overall convergence would again be $O(\frac{1}{n})$ if $\alpha \sim \delta = O(\frac{1}{n^2})$ and $\eta = O(\frac{1}{n^4})$.

As a specific numerical example we considered the estimation of a in (3.1) where

$$a_0 = 1 + \sin\pi x,\ u(a_0) = \sin 2\pi x,\ c = 1,$$
$$f = -(a_0 u(a_0)_x)_x + u(a_0) = 2\pi^2(2(1 + \sin\pi x)\sin 2\pi x - \cos\pi x\cos 2\pi x) + \sin 2\pi x.$$

For this choice of f and $u(a_0)$, a_0 is the unique element in $H^1(0,1)$ satisfying $u(a) = u(a_0)$. One can show that a function ω satisfying (3.2), (3.3) exists, provided that the estimator a^* has the form

$$a^*(\lambda, k) = a_0 + \lambda w, \quad \text{where} \quad \lambda \in \mathbf{R} \quad \text{and}$$

(3.5)
$$w = \frac{\cos((2-k)\pi x}{2(1+(2\pi - k\pi)^2)} + \frac{\cos((2+k)\pi x)}{2(1+(2\pi + k\pi)^2)}.$$

The perturbed data where generated in the following manner

$$z^\delta = P^n u(a_0) + \delta \sum_{i=0}^{n} r_i B_i^n,$$

where δ characterizes the error level, r_i are uniformly distributed random numbers in $[-1,1]$, B_i^n are the basis functions for Y_n and P^n denotes the orthogonal projection of $L^2(0,1)$ onto Y_n in $L^2(0,1)$. The finite dimensional problems were solved with the Levenberg-Marquardt algorithm that is available in the IMSL-library, and the solutions are denoted by $a_n = a_{\alpha(n),n}^{\delta(n)}$. We did not attempt to estimate the numerical error $\eta(n)$ involved in solving $(\mathcal{P})$. In all plots except for the last one, the abscissa gives the values for $\ln n$ and the ordinate those for $\ln |a_0 - a_n|_{H^1}$. Thus in the case where one expects $O(\frac{1}{n})$ respectively $O(\frac{1}{\sqrt{n}})$ convergence, the values for $\ln |a_0 - a_n|$ should lie approximately on a line with slope -1 respectively $-\frac{1}{2}$. Due to the fact that different random numbers are involved in defining the perturbed observations $y_{\delta(n)}$, one cannot expect that the values for $\ln |a_0 - a_n|_{H^1}$ lie on a perfectly straight line. The finite dimensional problems were solved for the values $n = 6, 8, \ldots, 36$.

In Plots 1 and 2 we give the results for $\alpha(n) = \frac{1}{100n}$ with $\delta = 0$ and $\delta = 10\alpha$ respectively. In each case the estimator was chosen according to (3.5) with $\lambda = -50$, $k = 3$, so that in particular (H3) is satisfied. The expected and observed error rate is $O(\frac{1}{\sqrt{n}})$ in either case. In Plot 3 the results are given for $\alpha(n) = \frac{1}{100n}$ and $\delta = 10\alpha$, with the estimator a^* now changed to be the function with constant value 1.5. The error rate is now much lower than $O(\frac{1}{\sqrt{n}})$.

For Plot 4 we took $\alpha(n) = \frac{1}{500n^2}$, $\delta = 100\alpha$, and the estimator was chosen according to (3.5) with $\lambda = -50$, $k = 3$. The expected and observed rate is $O(\frac{1}{n})$. If the estimator is replaced by $a^* = 2 - x$, while the other specifications remain fixed, then the rate of convergence is much lower than $O(\frac{1}{\sqrt{n}})$, see Plot 5. For Plot 6 we repeated this calculation with $\alpha(n) = \frac{1}{500n^2}$, $\delta = 100\alpha$, and used iterative estimator improvement, i.e. starting with the estimator $a^* = 2 - x$ for $n = 6$, we used as estimator for $n = 8, 10, \ldots$ the result a_{n-2} obtained in the previous step. For this procedure the convergence rate is again $O(\frac{1}{n})$. In the last plot the solid line shows a_0 and the dotted line gives the result a_{16} with the other specifications corresponding to those of Plot 4.

References.

[BK] H.T. Banks and K. Kunisch, Estimation Techniques for Distributed Parameter Systems, Birkhäuser, Boston, 1989.

[EKN] H.W. Engl, K. Kunisch and A. Neubauer, Tikhonov regularization for the solution of nonlinear ill–posed problems, Inverse Problems 5 (1989), 523–540.

[Ge] G. Geymayer, Regularisierungsverfahren und deren Anwendung auf Inverse Randwertprobleme, Diplomarbeit, Technical University of Graz, 1988.

[G] C.W. Groetsch, The Theory of Tikhonov Regularization for Fredholm Integral Equations of the First Kind, Pitman, Boston, 1984.

[GK] G. Geymayer and K. Kunisch, Convergence rates for regularized illposed problems, Technical Report at Technical University of Graz.

[KW] K. Kunisch and L. White, Regularity properties in parameter estimation of diffusion coefficients in elliptic boundary value problems, Appl. Analysis, 21 (1986), 71-87.

[M] V.A. Morozov, Methods for Solving Incorrectly Posed Problems, Springer Verlag, New York, 1984.

[Mo] V. Mosco, Convergence of convex sets and solutions of variational inequalities, Adv. Math. 3 (1969), 510–585.

[N] A. Neubauer, Tikhonov regularization for non-linear illposed problems: optimal convergence rates and finite dimensional approximation, Inverse Problems 5 (1989), 541–558.

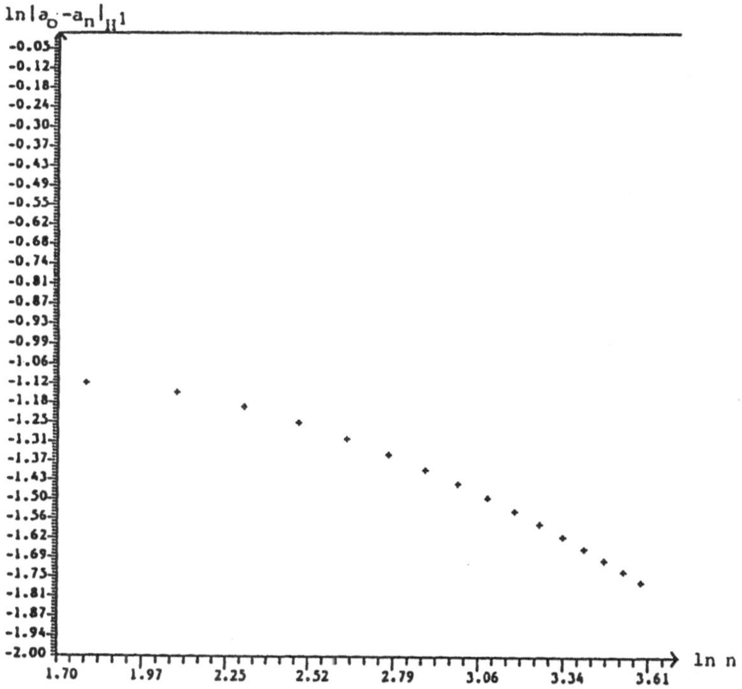

Plot 1

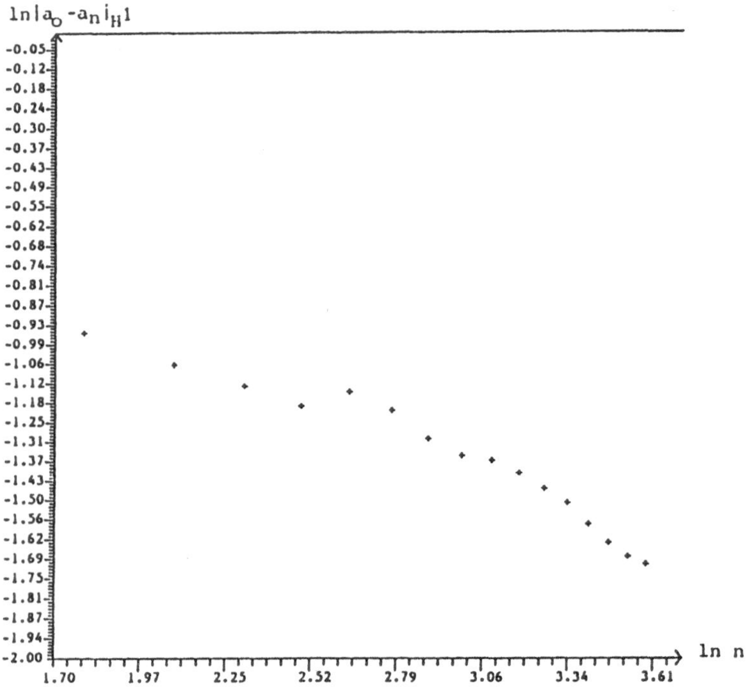
ln|a_o - a_n|_H|
-0.05
-0.12
-0.18
-0.24
-0.30
-0.37
-0.43
-0.49
-0.55
-0.62
-0.68
-0.74
-0.81
-0.87
-0.93
-0.99
-1.06
-1.12
-1.18
-1.25
-1.31
-1.37
-1.43
-1.50
-1.56
-1.62
-1.69
-1.75
-1.81
-1.87
-1.94
-2.00
1.70
1.97
2.25
2.52
2.79
3.06
3.34
3.61
ln n

Plot 2

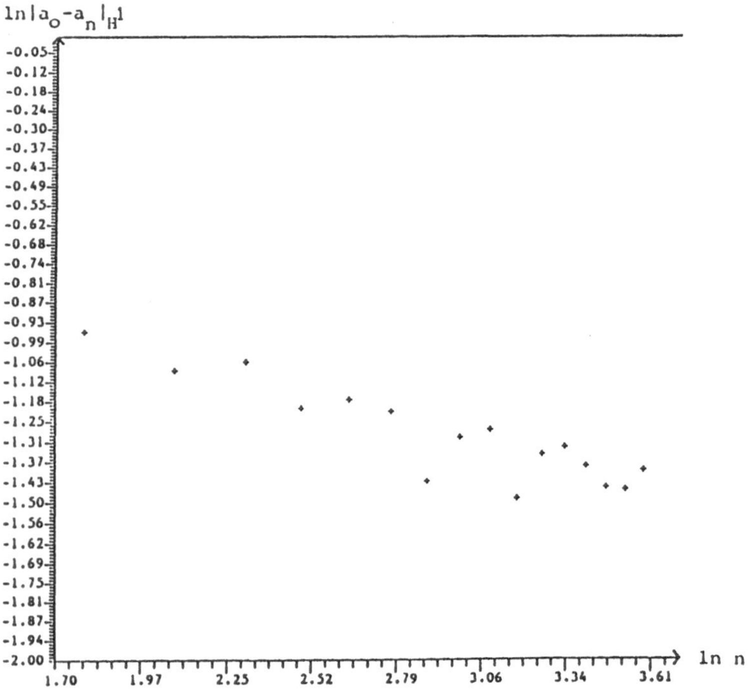
ln|a_o - a_n|_H|
-0.05
-0.12
-0.18
-0.24
-0.30
-0.37
-0.43
-0.49
-0.55
-0.62
-0.68
-0.74
-0.81
-0.87
-0.93
-0.99
-1.06
-1.12
-1.18
-1.25
-1.31
-1.37
-1.43
-1.50
-1.56
-1.62
-1.69
-1.75
-1.81
-1.87
-1.94
-2.00
1.70
1.97
2.25
2.52
2.79
3.06
3.34
3.61
ln n

Plot 3

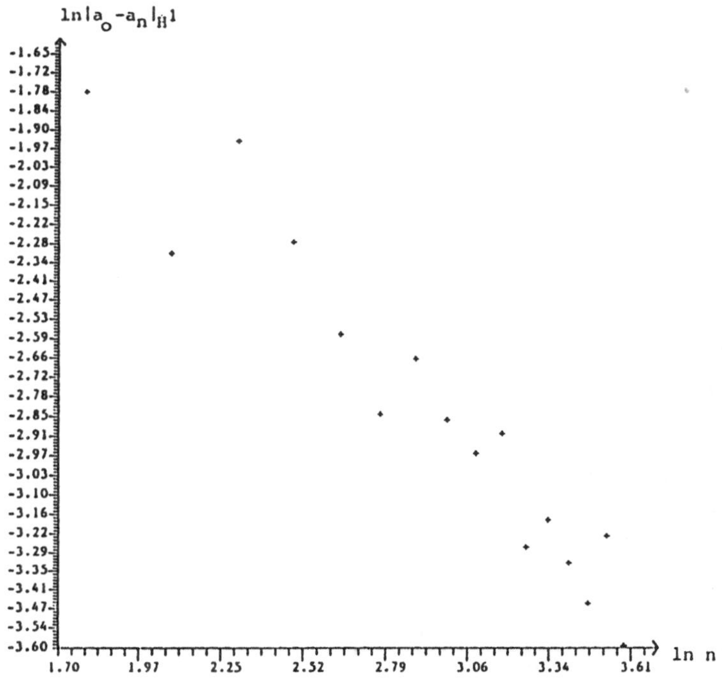

Plot 4

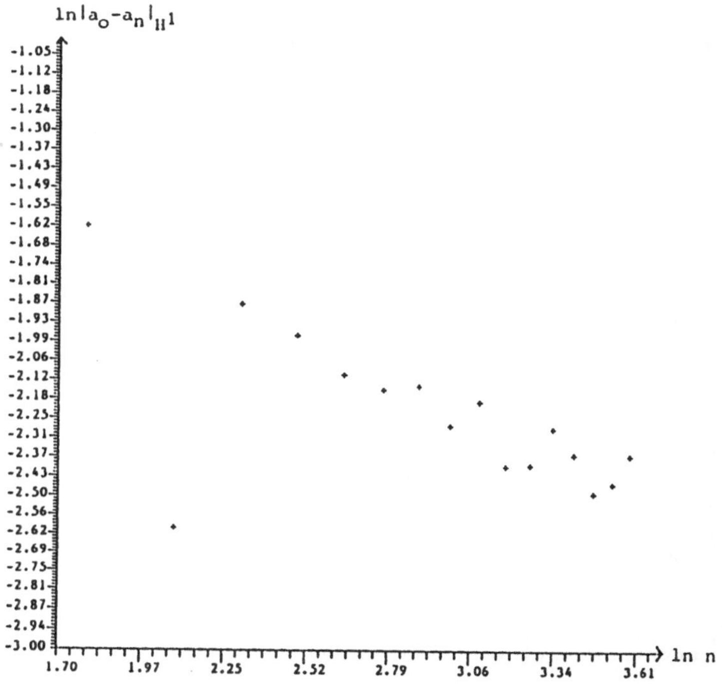

Plot 5

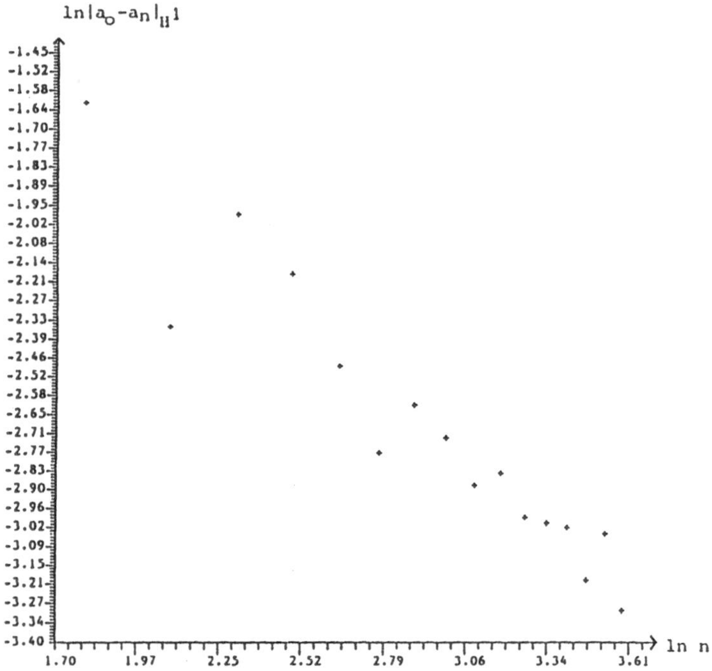
ln|a0-an|H1
-1.45
-1.52
-1.58
-1.64
-1.70
-1.77
-1.83
-1.89
-1.95
-2.02
-2.08
-2.14
-2.21
-2.27
-2.33
-2.39
-2.46
-2.52
-2.58
-2.65
-2.71
-2.77
-2.83
-2.90
-2.96
-3.02
-3.09
-3.15
-3.21
-3.27
-3.34
-3.40
1.70
1.97
2.25
2.52
2.79
3.06
3.34
3.61
ln n

Plot 6

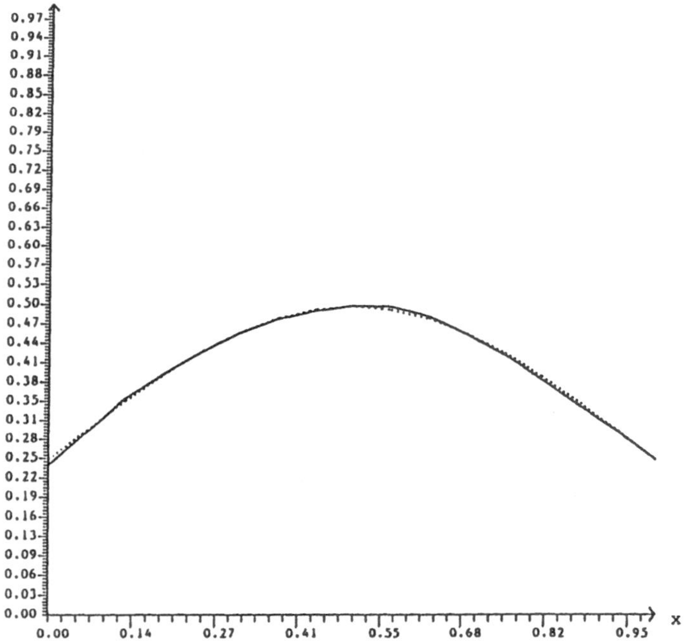
0.97
0.94
0.91
0.88
0.85
0.82
0.79
0.75
0.72
0.69
0.66
0.63
0.60
0.57
0.53
0.50
0.47
0.44
0.41
0.38
0.35
0.31
0.28
0.25
0.22
0.19
0.16
0.13
0.09
0.06
0.03
0.00
0.00
0.14
0.27
0.41
0.55
0.68
0.82
0.95
x

Plot 7

ON INVERSE PROBLEMS FOR EVOLUTIONARY SYSTEMS: GUARANTEED ESTIMATES AND REGULARIZED SOLUTIONS

A.B. Kurzhanski

International Institute for Applied Systems Analysis, A-2361 Laxenburg, Austria

I.F. Sivergina

Institute of Mathematics and Mechanics of the Ural Scientific Center
Academy of Sciences of the USSR, Sverdlovsk, USSR

This paper deals with the selection of an initial distribution in the first boundary-value problem for the heat equation in a given domain $[0,\theta] \times \Omega$, $\theta < \infty$ with zero values on its boundary S so that the deviation of the respective solution from a given distribution would not exceed a preassigned value $\gamma > 0$. The result is formulated here in terms of the "theory of guaranteed estimation" for noninvertible evolutionary systems. It also allows an interpretation in terms of regularization methods for ill-posed inverse problems and in particular, in terms of the quasiinvertibility techniques of J.-L. Lions and R. Lattes.

1. The Problem.

Assume Ω to be a compact domain in $\mathbf{R}^n$ with a smooth boundary S; $\theta > 0$, $\gamma > 0$ to be given numbers, functions $y(t,x)$, $z(x)(\mathbf{R} \times \mathbf{R}^n \to \mathbf{R}^1)$, $(\mathbf{R}^n \to \mathbf{R}^1)$ to be given and such that $y(\cdot,\cdot) \in L_2([0,\theta] \times \Omega)$, $z(\cdot) \in L_2(\Omega)$.

Denote $u = u(t,x;\ w(\cdot))$ to be the solution to the boundary value problem

$$\frac{\partial u}{\partial t} - \Delta u = 0\,, \quad 0 \le t \le \theta\,, \tag{1}$$

$$u|_{[0,\theta]\times S} = 0\,,$$

$$u|_{t=0} = w(\cdot)\,,$$

Also denote

$$J(w(\cdot)) = \alpha \int_0^\theta \int_\Omega (u(t,x;\ w(\cdot)) - y(t,x))^2 dx dt + \tag{2}$$

$$+ \beta \int_{\Omega} (u(\theta,x;\ w(\cdot)) - z(x))^2 dx$$

with $\alpha \geq 0$, $\beta \geq 0$.

Consider the following problem: among the possible initial distributions $w(\cdot) \in L_2(\Omega)$ specify a distribution $w^0(\cdot)$ that ensures

$$J(w^0(\cdot)) \leq \gamma\ . \tag{3}$$

The latter is an *inverse problem* [1]. With $\alpha = 0$ it was studied by J.-L. Lions and R. Lattes within the framework of the method of "quasiinvertibility" [2]. Numerical stability was ensured in this approach.

Let us now transform the previous problem into the following: among the distributions $w(\cdot) \in L_2(\Omega)$ determine *the set* $W^*(\cdot) = \{w^*(\cdot)\}$ *of all those distributions* $w^*(\cdot)$ *that yield the inequality*

$$J(w^*(\cdot)) \leq \gamma\ .$$

Assuming that the problem is solvable ($W^*(\cdot) \neq \phi$) we may describe its solution in terms of the theory of "guaranteed observation" [3]. Namely, assume $y(t,x)$, $z(x)$ to be the available measurements of the process (1), so that

$$y(t,x) = u(t,x;\ w(\cdot)) + \xi(t,x) \tag{4}$$

$$z(x) = u(\theta,x;\ w(\cdot)) + \sigma(x)$$

$$0 \leq t \leq \theta,\ \ x \in \Omega$$

where $\xi(t,x)$, $\sigma(x)$ stand for the *measurement noise* which is *unknown* in advance *but bounded* by the restriction

$$\alpha \int_0^{\theta} \int_{\Omega} \xi^2(t,x)\ dxdt + \beta \int_{\Omega} \sigma^2(x) dx \leq \gamma\ . \tag{5}$$

Then $W^*(\cdot)$ will be precisely the *set of all initial states* of system (1) *consistent with measurements* $y(t,x)$, $z(x)$ (4) *and with restriction* (5).

The aim of this paper will be to describe some stable schemes of calculating the sets $W^*(\cdot)$ and their specific elements. (A direct calculation of these may obviously lead to unstable numerical procedures.)

2. The Regularizing Problem (A General Solution)

Consider a rather general problem. Assume the values ξ, σ, w to be unknown in advance while satisfying a joint quadratic constraint

$$\langle w(\cdot),\, N(\varepsilon)w(\cdot)\rangle + \alpha \int_0^\theta \langle \xi(t,\cdot),\, M(\varepsilon)\xi(t,\cdot)\rangle\, dt \tag{6}$$

$$+ \beta\, \langle \sigma(\cdot),\, K(\varepsilon)\sigma(\cdot)\rangle \le \gamma + h_\varepsilon\,, \quad h_\varepsilon > 0\,,$$

Here $N(\varepsilon)$, $M(\varepsilon)$, $K(\varepsilon)$ are nonnegative self-adjoint operators from $\mathcal{L}_2(\Omega)$ into itself (with $N(\varepsilon)$ invertible) and such that each of them depends on a small parameter $\varepsilon>0$. The symbol $\langle\cdot,\cdot\rangle$ denotes a scalar product in $L_2(\Omega)$.

An *informational set* $W_\varepsilon(\cdot)$ of distributions $w(\cdot)$ consistent with measurements y and z will be defined as the variety of those and only those functions $w(\cdot) \in L_2(\Omega)$ for each of which there exists such a pair $\xi(\cdot,\cdot) \in \mathcal{L}_2([0,\theta] \times \Omega)$ and $\sigma(\cdot) \in \mathcal{L}_2(\Omega)$ that equalities (1), (4) would be fulfilled together with the inequality (6).

Lemma 2.1. The informational set $W_\varepsilon(\cdot)$ consists of all those functions $w(\cdot) \in \mathcal{L}_2(\Omega)$ that satisfy the inequality

$$\langle w(\cdot) - w_\varepsilon^0(\cdot),\, \mathbf{B}(\varepsilon)(w(\cdot) - w_\varepsilon^0(\cdot))\rangle \le \gamma + h_\varepsilon - \kappa_\varepsilon^2 \tag{7}$$

where

$$\mathbf{B}(\varepsilon) = N(\varepsilon) + U^*M(\varepsilon)U + U_\theta^*K(\varepsilon)U_\theta\,,$$

$$w_\varepsilon^0(\cdot) = \mathbf{B}^{-1}(\varepsilon)(U^*M(\varepsilon)y(\cdot,\cdot) + U_\theta^*K(\varepsilon)z(\cdot))\,,$$

$$\kappa_\varepsilon^2 = \langle z(\cdot),\, K(\varepsilon)z(\cdot)\rangle + \int_0^\theta \langle\, y(t,\cdot),\, M(\varepsilon)y(t,\cdot)\rangle\, dt\,,$$

$$(Uw(\cdot))(t,x) = u(t,x;\, w(\cdot))\,,\quad (U_\theta w(\cdot))(x) = u(\theta,x;\, w(\cdot))\,;$$

$$U : \mathcal{L}_2(\Omega) \to \mathcal{L}_2([0,\theta] \times \Omega);\ U_\theta : \mathcal{L}_2(\Omega) \to \mathcal{L}_2(\Omega)$$

and where U^* stands for the respective adjoint operator.

It is further assumed that h_ε is such that $W_\varepsilon(\cdot)$ is nonvoid.

If there exists an $\varepsilon_0 \ge 0$ such that

$$J(w_\varepsilon^0(\cdot)) \to \inf\{J(w(\cdot)) \mid w(\cdot) \in \mathcal{L}_2(\Omega)\}$$

$$\text{with } \varepsilon \to \varepsilon_0$$

then the problem of estimating the distributions $w(\cdot)$ due to the system (1), (4), (6) will be further referred to as the ***regularizing problem*** for problem (1), (3).

3. Quasiinvertibility

With $\alpha = 0$ in equation (2) we arrive at the problem investigated in [2] by means of the *quasiinvertibility* techniques. Following the latter consider an auxiliary boundary-value problem

$$\frac{\partial V_\varepsilon}{\partial t} - \Delta V_\varepsilon - \varepsilon \Delta^2 V_\varepsilon = 0 \ , \ \ 0 \le t \le \theta \ , \ \ (\varepsilon > 0)$$

$$V_\varepsilon|_{[0,\theta] \times S} = \Delta V_\varepsilon|_{[0,\theta]\times S} = 0$$

$$V_\varepsilon|_{t=\theta} = z(\cdot) \ .$$

Then taking

$$w_\varepsilon(\cdot) = V_\varepsilon(0,\cdot) \tag{8}$$

we come to

$$\mathcal{J}(w_\varepsilon(\cdot)) \to 0 \quad (\varepsilon \to 0) \ .$$

The following question does arise: is it possible to select the operators $N(\varepsilon)$, $M(\varepsilon)$, $K(\varepsilon)$ that define the quadratic constraint (6) in such a way that the center $w_\varepsilon^0(\cdot)$ of the informational ellipsoid $W_\varepsilon(\cdot)$ would coincide with the solution $V_\varepsilon(0,\cdot)$ of Lions and Lattes?

Assume $0 \le \lambda_1 \le \lambda_2 \le \cdots \le \lambda_i \cdots$ to be the eigenvalues and $\{\varphi_i(\cdot)\}$ to be the respective complete system of orthonormal eigenfunctions in the first boundary-value problem for the operator $A = -\Delta$ in the domain Ω.

Assume

$$(N(\varepsilon)w)(\cdot) = \sum_{i=1}^{\infty} (1 - e^{\varepsilon\lambda_i^2\theta}) w_i \varphi_i(\cdot) \tag{9}$$

$$(K(\varepsilon)\sigma)(\cdot) = \sum_{i=1}^{\infty} e^{-(\varepsilon\lambda_i^2 - 2\lambda_i)\theta} \sigma_i \varphi_i(\cdot)$$

with w_i (respectively σ_i, z_i) being the Fourier coefficients for the expansion of functions $w(\cdot)$ (respectively $\sigma(\cdot)$, $z(\cdot)$) in a series along the system of functions $\{\varphi_i(\cdot)\}$.

Theorem 3.1. Assume $\alpha = 0$ and operators $N(\varepsilon)$, $M(\varepsilon)$, $K(\varepsilon)$ of inequality (6) to be defined as in (9) with $M(\varepsilon) = 0$. Then for all $\varepsilon > 0$ the center $w^0(\cdot)$ of the ellipsoid $W_\varepsilon(\cdot)$ (7) will coincide with the "Lions - Lattes" solution $w_\varepsilon(\cdot)$ (8). Namely

$$w_\varepsilon^0(\cdot) = w_\varepsilon(\cdot) = V_\varepsilon(0,\cdot)$$

and $w_\varepsilon^0(\cdot)$ will be represented as

$$w_\varepsilon^0(\cdot) = \sum_{i=1}^{\infty} e^{(-\varepsilon\lambda_i^2+\lambda_i)\theta} z_i\varphi_i(\cdot) \ .$$

The next theorem indicates that an appropriate selection of the operators $N(\varepsilon)$, $K(\varepsilon)$ in (6) (with $M(\varepsilon) = 0$) would allow to approximate the set

$$W^*(\cdot) = \{w^*(\cdot) | J(w^*(\cdot)) \le \gamma\}$$

with respective informational sets $W_\varepsilon(\cdot)$.

Theorem 3.2. Assume $\alpha = 0$, $\beta = 1$, $\varepsilon > 0$, $\nu > 0$ and the operators $N(\varepsilon)$, $M(\varepsilon)$, $K(\varepsilon)$ of inequality (6) to be defined as

$$(N(\varepsilon)w)(\cdot) = (N_{\varepsilon,\nu}w)(\cdot) =$$

$$\sum_{i=1}^{\infty} \left(e^{-2(1+\nu\lambda_i)^{-1}\lambda_i\theta} - e^{-(\varepsilon\lambda_i^2+2\lambda_i)\theta}\right) w_i\varphi_i(\cdot) \ ,$$

$$(K(\varepsilon)\sigma)(\cdot) = (K_\varepsilon\sigma)(\cdot) = \sum_{i=1}^{\infty} e^{-\varepsilon\lambda_i^2\theta} \sigma_i\varphi_i(\cdot) \ ,$$

$$M(\varepsilon) = 0 \ .$$

Then with $h_\varepsilon = 0$ there exists a pair $\varepsilon_0 > 0$, $\nu_0 > 0$ such that with $\varepsilon \le \varepsilon_0$, $\nu \le \nu_0$ the respective informational ellipsoidal set $W_\varepsilon(\cdot) = W_{\varepsilon,\nu}(\cdot) \ne \phi$. Its centers $w_{\varepsilon,\nu}^0$ converge:

$$\lim w_{\varepsilon,\nu}^0 = w_\varepsilon(\cdot) \quad (\nu \to 0)$$

and

$$\lim W_{\varepsilon,\nu}(\cdot) = W^*(\cdot) \quad (\varepsilon \to 0, \nu \to 0)$$

in the sense of Kuratowski [4].

4. Extremality and the General Regularization Scheme

Consider the minimization process for the functional (2). With $\alpha = 0$ a numerically stable scheme for calculating $\inf J$ is ensured by the quasiinvertibility method discussed above. We will now proceed with the construction of a respective algorithm for the general case, particularly for $\beta \geq 0$.

Theorem 4.1. The value

$$\inf_{w(\cdot)} J = \alpha \int_0^\theta \|y(t)\|^2\, dt + \beta\|z(\cdot)\|^2 - \sum_{i=1}^{\infty} v_i(\alpha p_i + \beta\, e^{-\lambda_i \theta}\, z_i)^2 ,$$

where

$$v_i = 2\lambda_i \left[2\lambda_i \beta\, e^{-2\lambda_i \theta} + \frac{\alpha(1 - e^{-2\lambda_i \theta})}{2\lambda_i} \right]^{-1}$$

$y_i(t)$, p_i are the Fourier coefficients for $y(t,\cdot)$, $p(\cdot)$,

$$p(x) = \int_0^\theta u(t,x;y(t,x))\, dt$$

$$y(t) = \{y_1(t),\ldots,y_k(t),\ldots\} ,$$

is a sequence in ℓ_2. The sequence

$$w_\varepsilon(\cdot) = \sum_{i=1}^{\infty} v_i(\alpha\, e^{-\varepsilon\lambda_i}\, p_i + \beta\, e^{-(\varepsilon\lambda_i^2 + \lambda_i)\theta}\, z_i)\varphi_i(\cdot) \tag{10}$$

minimizes $J(w(\cdot))$ with $\varepsilon \to 0$.

Theorem 4.2. Suppose $\beta = 0$. Then for $w_\varepsilon(\cdot)$ of (10) we will have

$$w_\varepsilon(x) = -2(\Delta u(\varepsilon,x;p(\cdot)) + \sum_{k=1}^{\infty} \Delta u(2\theta k,x;p(\cdot)))$$

and consequently

$$J(w_\varepsilon(\cdot)) \to \inf_{w(\cdot)} J(w(\cdot)) \quad \textit{with} \quad \varepsilon \to 0 .$$

Remark 4.1. Once there exists a distribution $w(\cdot) \in L_2(\Omega)$ that ensures the equalities

$$y(t,x) \equiv u(t,x;w(\cdot))$$

$$z(x) \equiv u(\theta,x;w(\cdot))$$

the value

$$\inf_{w(\cdot)} \mathcal{J}(w(\cdot)) = 0 \ .$$

The next question is whether the functions $w_\varepsilon(\cdot)$ of (10) could serve as centers of some "informational ellipsoids" $\mathbf{W}_\varepsilon$ that would correspond to an appropriate selection of operators $N(\varepsilon)$, $M(\varepsilon)$, $K(\varepsilon)$ in the restriction (6). The answer is affirmative and is given by the following theorem.

Theorem 4.3. Suppose the restriction (6) is defined through the operators

$$(M(\varepsilon)\xi)(t,x) = 2 \sum_{i=1}^{\infty} e^{-\varepsilon\lambda_i} \lambda_i (1 - e^{-2\lambda_i \theta})^{-1} \xi_i(t) \varphi_i(x) \tag{11}$$

with $N(\varepsilon)$, $K(\varepsilon)$ being the same as in (9). *Then the center $w_\varepsilon^0(\cdot)$ of the respective informational domain $\mathbf{W}_\varepsilon$ for equation* (1) *under restriction* (6), (9), (11) *will coincide with the distribution given by formula* (10): $w_\varepsilon^0(\cdot) = w_\varepsilon(\cdot)$.

Remark 4.2. Define a *minmax estimate* w^0 for a bounded convex set $\mathbf{W}$ as its *Chebyshev center*:

$$\sup\{\|w^0 - w\| \mid w \in \mathbf{W}\} = \min_{z \in \mathbf{W}} \sup \{\|z - w\| \mid w \in \mathbf{W}\} \ .$$

Then once $\mathbf{W}$ is an ellipsoid its Chebyshev center w^0 will coincide with its formal center. For an arbitrary bounded informational set that may appear in nonlinear nonconvex problems its Chebyshev center may be taken as a natural "guaranteed estimate" for the unknown parameter w.

5. Other Regularizing Procedures

Consider $\alpha = 0$. (a) Another regularizing procedure may be designed through the solution $v_\varepsilon(t,x)$ to the following problem:

$$\frac{\partial}{\partial t}(v_\varepsilon - \varepsilon \Delta v_\varepsilon) - \Delta v_\varepsilon = 0, \ \ 0 \le t \le \theta$$

$$v_\varepsilon \left.\right|_{[0,\theta]\times S} = 0, \quad v_\varepsilon \left.\right|_{t=\theta} = z(\cdot)$$

so that

$$w_\varepsilon(\cdot) = v_\varepsilon(0,\cdot) \ . \tag{12}$$

The system (12) was introduced in paper [5]. The function $w_\varepsilon(\cdot) = v_\varepsilon(0,\cdot)$ will be the center of the respective informational ellipsoid consistent with measurement $z(\cdot)$ if we assume

$$(N(\varepsilon)w)(\cdot) = \sum_{i=1}^{\infty} e^{-\lambda_i\theta}(1 - e^{-\varepsilon(1+\varepsilon\lambda_i)^{-1}\lambda_i^2\theta})w_i\varphi_i(\cdot)$$

$$(K(\varepsilon)\sigma)(\cdot) = \sum_{i=1}^{\infty} e^{(1+\varepsilon\lambda_i)^{-1}\lambda_i\theta}\sigma_i\varphi_i(\cdot)\ ,\quad M(\varepsilon) = 0\ .$$

Here the center of the ellipsoid is defined in a formal way, through formula (7). The ellipsoid itself is however unbounded.

(b) With $z(\cdot)$ given, assume that there exists a solution to equation

$$U_\theta w(\cdot) = z(\cdot)$$

Consider the constraint (6) with

$$(N(\varepsilon)w)(\cdot) = n_\varepsilon w(\cdot),\ (K(\varepsilon)\sigma)(\cdot) = k_\varepsilon\sigma(\cdot)\ ,\quad M(\varepsilon) = 0$$

where $n_\varepsilon > 0,\ k_\varepsilon > 0$ are real numbers.

Then with $n_\varepsilon = \varepsilon^2$, $k_\varepsilon = 1$ the center $w_\varepsilon^0(\cdot)$ of the respective ellipsoid $W_\varepsilon(\cdot)$ will coincide with the quasisolution (in the sense of V.K. Ivanov [6]) to the equation

$$U_\theta w(\cdot) = z(\cdot)\ ,$$

on the set

$$M = \{w(\cdot) \mid \|w(\cdot)\| \leq \|w_\varepsilon^0(\cdot)\|\}\ ,\ \text{i. e.}$$

$$w_\varepsilon^0(\cdot) = \arg\min \|U_\theta w(\cdot) - z(\cdot)\|\ ,\quad w(\cdot) \in M\ .$$

(c) Assuming $n_\varepsilon = 1$, $k_\varepsilon = \varepsilon^{-2}$ the function $w_\varepsilon^0(\cdot)$ will be an approximate solution to the equation

$$U_\theta w(\cdot) = z(\cdot)$$

by the "bias method" with bias

$$d(U_\theta w(\cdot), z(\cdot)) = J(w(\cdot))\ .$$

So that

$w_\varepsilon^0(\cdot)$ would solve the problem

$$\min \{\|w(\cdot)\| : d(U_\theta w(\cdot),\, z(\cdot)) \le J(w_\varepsilon^0(\cdot))\}$$

In both cases (b), (c) we observe that $J(w_\varepsilon^0(\cdot)) \to 0$ with $\varepsilon \to 0$.

6. A Continuity Theorem

Taking the solution (10) present it as a linear maping

$$w_\varepsilon(\cdot) = F_\varepsilon(y(\cdot,\cdot), z(\cdot))$$

from $L_2([0,\theta]\times\Omega) \times L_2(\Omega)$ into $L_2(\Omega)$.

Suppose

$$y_\delta(t,x) = u(t,x;\, w^*(\cdot)) + \xi_\delta(t,x)$$

$$z_\delta(x) = u(\theta,x;\, w^*(\cdot)) + \sigma_\delta(x)$$

where

$$\|\xi_\delta(\cdot,\cdot)\| \le \delta_1, \quad \|\sigma_\delta(\cdot)\| \le \delta_2;\ \ \delta_1,\delta_2 > 0\ .$$

Theorem 6.1. The mapping F_ε is uniformly continuous in $L_2([0,\theta]\times\Omega) \times L_2(\Omega)$. The following estimate is true

$$\|F_\varepsilon(y_\delta(\cdot,\cdot),\, z_\delta(\cdot)) - w^*(\cdot)\| \le \left[R(\varepsilon,w^*(\cdot)) + \frac{\delta_1 c}{\sqrt{\varepsilon}} + \delta_2 e^{\frac{\theta}{4\varepsilon}}\right], \quad R(\varepsilon,\omega^*(\cdot)) = \left[\sum_{i=1}^{\infty} w_i^{*2}(1-e^{-\varepsilon\lambda_i^2\theta})^2\right]^{1/2}.$$

With $\varepsilon \to 0$, $(\delta_1 / \sqrt{\varepsilon}) \to 0$, $(\delta_2\, e^{\frac{\theta}{4\varepsilon}}) \to 0$, there is a strong convergence $F_\varepsilon(y_\delta(\cdot,\cdot),\, z_\delta(\cdot)) \to w^*(\cdot)$.

References

[1] Tikhonov A.N., Arsenin V.Ya. Methods of Solving Ill-posed Problems. Nauka, Moscow, 1986.

[2] Lions J.-L., Lattes R. Méthode de Quasi-Réversibilité et Applications. Dunod, Paris, 1967.

[3] Kurzhanski A.B. Control and Observation Under Uncertainty. Nauka, Moscow, 1977.

[4] Kuratowski R. Topology Vol. 1, 2. Academic Press, 1966.

[5] Gaewski H., Zacharias K. Zur Regularisierung einer Klasse nichtkorrekter Probleme bei Evolutionsgleichungen. J. Math. Anal. & Appl., V. 38, No. 3, 1972.

[6] Ivanov V.K., Vasin V.V., Tanana V.P. The Theory of Linear Ill-posed Problems and Its Applications. Nauka, Moscow, 1978.

EXPONENTIAL STABILIZATION, VIA RICCATI OPERATOR, OF HYPERBOLIC SYSTEMS WITH UNCONTROLLED, UNBOUNDED PERTURBATIONS

I. Lasiecka*

Department of Applied Mathematics
University of Virginia
Charlottesville, VA 22903

1. Introduction

Let A be a generator of a s.c. semigroup e^{At} on a Hilbert space H with domain D(A). Let $[D(A)]'$ denote the dual (pivotal) space to D(A) with respect to the H-inner product. D(A) is equipped, as usual, with a graph norm topology and $[D(A)]'$ is equipped with a norm given by $|u|_{[D(A)]'} = |[A^* - \lambda_0]^{-1} u|_H$ where $\lambda_0 \in \rho(A^*)$. Without loss of generality, we shall assume $\lambda_0 = 0$.

We shall introduce the following operators:

(1.1) The linear (generally unbounded on H) operators

$B_i \in \mathcal{L}(U_i \to [D(A^*)]')$; $i = 1, 2$ where U_i denote another Hilbert space. The operators B_i are required to satisfy

(H-1) (i) For some $T > 0$ there exists constant C_T such that

$$\int_0^T |B_2^* e^{A^*t} x|_{U_2}^2 \, dt \le C_T |x|_H^2; \quad x \in D(A^*).$$

(ii) $$D(B_2^*) \subseteq D(B_1^*).$$

(1.2) The nonlinear, continuously Frechet differentiable operator

$$G: H \to U_1 \quad \text{such that}$$

(H.2) $G(0) = 0$; $G'(0) = 0$

where $G'(y)$ stands for the Frechet derivative of $G(y)$.

(1.3) The linear densely defined operator

$$F: H \to U_2 \quad \text{such that} \quad D(A) \subset D(F).$$

Consider the following abstract model

(*)Research partially supported by the NSF Grant DMS-8301668 and by the AFOSR Grant AFOSR 89-0511.

$$(1.4)\quad \begin{cases} y_t = Ay + B_1\, G(y) + B_2 Fy \quad \text{on } [D(A^*)]' \\ y(0) = y_0 \in H\,. \end{cases}$$

The main goal of this paper can be formulated as follows: given the operators A, B_i, find the operator F such that the system described by (1.4) is locally exponentially stable for all nonlinear perturbations G subject to the condition (H-2). More precisely, we seek a stabilizing feedback F (depending only on A and B) such that the solutions y(t) of (1.4) corresponding to the initial data $y_0 \in B(0, R)$ (where B(0, R) denotes a ball in H with a radius R) decay exponentially to zero for all perturbations G taken from the class described by (H-2).

It should be noted that the main technical difficulties of this problem stem from the fact that the operators B_i are generally unbounded from $H \to U_1$. Indeed, in the case of ordinary differential equations or more generally, abstract differential equations with input operators B_i bounded (from $H \to U_1$), it is well known that any operator F stabilizing the linear part of the system, will produce local exponential stability of the nonlinearly perturbed system. This is not the case when the input operators B_i are unbounded and A is a generator at an arbitrary C_0 semigroup. In fact, it is known [see [L-T.1], [T.1] that the presence of unbounded perturbation B_1 G(y) (even if G is linear) may destroy the generation of the feedback semigroup, let-alone the exponential stability. Thus, the addition of unbounded nonlinear perturbation B_1 G(y) to the wellposed, stable system may, in general, destroy the desirable properties of the dynamics. In order to obtain the sought after stability results, special care must be given to the selection of the operator F. In this paper, we shall prove that under some additional restrictions placed on the system, the sought after stabilizing feedback F can be constructed via the solution to the Algebraic Riccati Equation. We shall establish that for this class of systems, the linear feedbacks given by the Riccati operator produce a robust stabilization in presence of uncontrolled nonlinear and unbounded perturbations.

We remark that our abstract model (1.4) incorporating nonlinear unbounded perturbation is motivated by several applications, (described in section 4) arising in control problems for the plate and wave equations with nonlinearly perturbed boundary conditions. Here, the effect of uncontrolled nonlinearities on the boundary is inherently unbounded and must be described by the unbounded operators B_i.

The outline of the paper is as follows. In section 2, we first recall some recent pertinent results on solvability of Riccati Equations with unbounded coefficients and then we shall formulate our main abstract theorem. The proof of the Main Theorem is relegated to section 3. Section 4 is devoted to concrete applications of our abstract results. As an example, we prove local exponential stabilizability of a Kirchhoff plate with nonlinear perturbations on the boundary.

2. Statement of the results

In order to formulate our results we need to introduce the following Algebraic Riccati Equation

(ARE) $(PAx, y)_H + (A^* Px, y)_H + (x, y)_H = (B_2^* Px, B_2^* Py)_U \; ; \; x, y \in D(A)$

If $B_2 \in \mathcal{L}(U_2, H)$ and the pair (A, B_2) is controllable or stabilizable, then the standard result [see [B-1])is that there exists positive selfadjoint solutions to (ARE). Recently, the above result has been extended to the case when the operator B_2 is unbounded, but subject to the hypothesis (H-1). In fact, the following result is available.

Theorem 1.1 [F-L-T]

Assume that the operator B_2 satisfies (H-1)(i). Assume moreover that the following "Finite Cost Condition" is satisfied

(F.C.C)

For any $y_0 \in H$, $\exists\, u \in L_2\,[0, \infty\,; U]$ such that

$$J(u, y) \equiv \int_0^\infty |y(t)|_H^2 + |u(t)|_U^2 \, dt < \infty$$

where y(u) satisfies

$$y_t = Ay + B_2 u; \;\; y(0) = y_0 \in H\,.$$

Then:

(i) There exists solution $P \in \mathcal{L}(H)$ to (ARE) such that $P \geq 0$; $P = P^*$. Moreover P enjoys the following regularity properties

(1.5) $P \in \mathcal{L}(D(A), D(A_p^*))$ where

$$A_p \equiv A - B_2 B_2^* P \text{ is a generator at a s.c.semigroup } e^{A_p t} \text{ on H.}$$

(1.6) $P \in \mathcal{L}(D(A_p), D(A^*)) \cap \mathcal{L}(D(A), D(A_p^*))$

(1.7) $B_2^* P \in \mathcal{L}(D(A), U_2) \cap \mathcal{L}(D(A_p); U_2)$.

(ii) the solution P is unique within the class of linear operators $P \in \mathcal{L}(H)$ such that $B_2^* P \in \mathcal{L}(D(A_p); U_2) \cap \mathcal{L}(D(A); U_2)$

(iii) $e^{A_p t}$ is exponentially stable on H i.e:

$$|e^{A_p t}|_{\mathcal{L}(H)} \leq Ce^{-\omega_0 t} \text{ for some } \omega_0 > 0\,.$$

Remark:

Notice that the above theorem, while it provides the existence of a bounded Riccati operator P, does not state that the gain operator $B_2^* P$ is bounded (in contrast with the classical results). This is a distinctive feature of the problem under study. We shall see later that, in general, when the operator B_2 is unbounded, the gain operator must remain unbounded.

Before we formulate our result we recall the following definitions.

Definition 2.1 the pair (A,B) is stabilizable on a Hilbert space H iff: there exist an operator F: $H \to U$, with $D(F) \supset D(A)$ such that $A_F \equiv A + BF$ generates a s.c and exponentially stable semigroup on H i.e: there exist constant $C > 0$; $\omega > 0$ such that

$$|e^{A_F t}|_{\mathcal{L}(H)} \leq C e^{-\omega t};\ t > 0.$$

Definition 2.2 the pair (A, B) is exactly controllable on a Hilbert space H for some $T_0 > 0$ iff for any $x_T \in H$ there exists $u \in L_2(0,T; U)$ such that $\int_0^T e^{A(T-\tau)} B u(\tau)\, d\tau = x_T$. The necessary and sufficient condition is that the following inequality holds:

$$\int_0^{T_0} |B^* e^{A^* t} x|_U^2\, dt \geq C_{T_0} |x|_H^2 .$$

Definition 2.3

System (1.4) is locally exponentially stable iff: given A, B and F subject to (H-1), (1.3), there exist constants $C > 0$; $\omega > 0$ such that solutions y of (1.4) with $\|y_0\|_H \leq R$ and with any perturbation G subject to the condition (H-2) satisfy

$$\|y(t)\|_H \leq C e^{-\omega t} \|y_0\|_H;\ t > 0$$

Now we are ready to formulate the main result of the paper.

Main Theorem

With reference to the system (1.4) assume that the hypothesis (H-1) and (H-2) are satisfied. Assume moreover that the pair

(1.8) (A, B_2) is either stabilizable or exactly controllable for some $T_0 > 0$.

(1.9) (A^*, I) is exactly controllable on H for some $T_0 > 0$.

Then the solutions y of (1.4) with the feedback operator, F given by

$$Fy = -B_2^* Py$$

(where P is the solution to (ARE)), are locally exponentially stable on H ■

Since in the case of a unitary group, the condition (1.9) is automatically satisfied we obtain

Collorary:

Assume that A generates an unitary group, the hypothesis (H-1) and (H-2) are satisfied and that (1.8) hold. Then the assertion of the Main Theorem holds true.

3. Proof of Main Theorem

The following two results will be crucially used in the proof of the Main Theorem.

Theorem 31. [L.1]

With reference to the system (1.4) we assume that the hypothesis (H-1), (1.3) and (H-2) are in force. Moreover we assume that

(A-1) $A + B_2F$ generates an exponentially stable semigroup on H.

(A-2) $(A + B_2 F)^{-1} B_1 \in \mathcal{L}(U_1; H)$

(A-3) There exist $T > 0$; $C_T > 0$ such that

$$\int_0^T |B_1^* e^{(A+B_2 F)t} x|_{U_1}^2 \, dt \le C_T \, |x|_H^2 .$$

Then the semilinear system (1.4) is locally exponentially stable on H.

Theorem 3.2 [F-L-T]

Assume the hypothesis of Theorem 1.1. In addition, assume that

(3.1) the pair (A^*, I) is controllable for some $T_0 > 0$.

Then the solution P to (ARE) is an isomorphism and $P^{-1} \in \mathcal{L}(H)$.

Remark 3.1 Notice that (3.1) holds whenever e^{At} is a group.

Remark 3.2 If $P^{-1} \in \mathcal{L}(H)$ then the gain operator $B_2^*P \in \mathcal{L}(H)$ iff $B_2 \in \mathcal{L}(U_2; H)$. Thus, in the case of a group, the gain operator B_2^*P must be unbounded if B_2 is unbounded. This fact should be contrasted with the case when e^{At} generates an analytic semigroup and the resulting gain operator B^*P is bounded even if B is unbounded (see [L-T.1, [F.1]).

Going back to the proof of our result, we notice first that by virtue of (1.8), the (F.C.C) conditions is automatically satisfied. Thus Theorem 1.1 yields the existence and uniqueness of the Riccati operator P. By (1.7) of Theorem 1.1 we also have: $B_2^*P \in \mathcal{L}(D(A); U_2)$, hence the feedback operator F defined by

(3.2) $Fy \equiv -B_2^*P\, y$

is densely defined on H, so (1.3) is verified. Therefore, the conclusion of our Main Theorem will follow from Theorem 3.1 once we establish validity of the assumptions (A.1) - (A.3). As for assumption (A.1) this, again, is a consequence of Theorem 1.1 part (iii). Indeed, by (3.2) $A_F \equiv A + B_2F = A - B_2 B_2^*P = A_P$.

In order to establish (A.2) and (A.3) we shall need several supporting Lemmas and Propositions.

Let P be the solution of (ARE). Define

(3.3) $D \equiv \text{Range } P|_{D(A)}$

By the assumption (1.9) and by Theorem 3.2 we have

(3.4) $P: H \to H$ is an isomorphism.

Now by density of D(A), we conclude that D is dense in H, and by (1.6) in Theorem 1.1 that $D \subset D(A_P^*)$. We shall prove that D is dense in the topology of $D(A_P^*)$.

Proposition 3.1

(3.5) D is dense in $D(A_P^*)$

(3.6) P^{-1} satisfies the following "Dual Riccati Equation".

(DRE) $(A P^{-1} x, y)_H + (A^* x, P^{-1} y)_H + (P^{-1} x, P^{-1} y)_H = (B_2^* x, B_2^* y)$ for $x, y \subset D \subset D(A_P^*)$.

Proof:

(i) proof of (3.5)

Let $z \in D(A_p^*)$ be such that

(3.7) $(d, z)_{D(A_P^*)} = 0$ for all $d \in D$.

We need to show that $z = 0$. (3.7) can be rewritten as $(d, A_P A_P^* z)_H = 0$ and since $d \in D$,

(3.7′) $(Px, A_P A_{P^*} z)_H = 0$ for all $x \in D(A)$.

Now, by (1.6) in Theorem 1.1 and by duality

$P \in \mathcal{L}(D(A_P^*)'; D(A)')$, hence

$P A_P A_P^* z \in D(A)'$ and

$P A_P A_P^* z \in D(A)'$ and

(3.8) $A^{*-1} P A_P A_P^* z \in H$;

From (3.7′)

(3.9) $(Ax, A^{*-1} P A_P A_P^* z)_H = 0$ for all $x \in D(A)$.

Since the Range of A is dense in H, by combining (3.8) and (3.9) we conclude

$A^{*-1} P A_P A_P^* z = 0$

or equivalently

$$P A_P A_P^* z = 0.$$

By (3.4) we know that P is injective on H. We shall show that P is also injective on a larger space; $[D(A_P^*)]'$. Indeed, it is enough to show that with any $\bar{x} \in H$

(3.10) $\quad P A_P \bar{x} = 0 \Rightarrow \bar{x} = 0.$

But from the explicit representation formula for the Riccati operator (see [F-L-T]) we obtain

$$P \hat{A}_p x = \int_0^\infty e^{\hat{A}^* \tau} [I + 2w P] e^{\hat{A}_P(\tau)} \hat{A}_p x \, d\tau = -\hat{A}^* Px - (I + 2w P)x \quad \text{on } D(A)'; \; x \in D(A_p)$$

where $\hat{A} = A - wI$; $\hat{A}_P \equiv A_P - wI$ and w is selected such that $|e^{\hat{A}t}|_{\mathcal{L}(H)} \le C e^{-t}$.

Thus

$$P A_P x = -A^* P x - (I + 2wP)x \quad \text{on}[D(A)]'$$

and $P A_P \bar{x} = 0$ implies that

(3.11) $\quad A^* P\bar{x} = -\bar{x} - 2w P\bar{x}$; and $\bar{x} \in D(A^* P)$.

Rewriting (ARE) as

$$(A_P x, Py) + (x, y) = -(A^* Px, y)$$

and recalling that $P^{-1} \in \mathcal{L}(H)$ we obtain

$$|A_P x|_H \le C[|x|_H + |A^* P x|_H]; x \in D(A^* P).$$

The above inequality shows that $D(A^* P) \subset D(A_P)$. Combining with (3.11) we conclude the following implication: $x \in H$ and $P A_P x = 0 \Rightarrow x \in D(A_P)$. Thus if $P A_P x = 0$ then $A_P x = 0$ follows from the injectivity of P on H. Since $e^{A_P t}$ is an exponentially stable semigroup, $0 \notin \sigma(A_P)$, x must be zero, proving (3.10). Thus $P A_P A_P^* z = 0$ implies that $A_P^* z = 0$ and by invertibility at A_P^*, $z = 0$ as desired.

<u>Proof of (3.6)</u>

Let $\bar{x} \equiv P^{-1} x$ and $\bar{y} \equiv P^{-1}$ where $x, y \in D$. Then $\bar{x}, \bar{y} \in D(A)$ and applying (ARE) with above $\bar{x}, \bar{y}$ yields

$$(A P^{-1} x, P P^{-1} y)_H + (A^* P P^{-1} x, P^{-1} y)_H + (P^{-1} x, P^{-1} y)_H = \langle B_2^* x, B_2^* y \rangle_{U_2} \; x, y \in D \subset \mathcal{D}(A_P^*).$$

this yields (3.6).

Next result asserts that Dual Riccati Equation holds on a larger space, namely $\mathcal{D}(A_P^*)$.

Lemma 3.1

The Dual Riccati Equation (DRE) is satisfied for all x, y ∈ D (A_P^*),

proof of Lemma 31. will follow through a seqnence of Propositions.

Proposition 3.2

$$P^{-1} \in \mathcal{L}(D(A_P^*); D(A)).$$

proof

We shall use (DRE) with x, y ∈ D. Since D ∈ D (A_P^*) we can write

$$(A P^{-1} x, y)_H + ([A^* - P B_2 B_2^*]x, P^{-1}y)_H + (P^{-1}x, P^{-1}y)_H = 0$$

or equivalently

(3.12) $\quad (A P^{-1}x, y)_H + (A_P^* x, P^{-1}y)_H + (P^{-1}x, P^{-1}y)_H = 0$

By the result (3.5) of Proposition 3.1, for any $x \in D\,(A_P^*)$ we can take $x_n \in D$ and $x_n \to x \in D(A_P^*)$ where the convergence is in $D(A_P^*)$ norm. Applying (3.12) with $x = x_n$ yields

$$(A P^{-1}x_n, y)_H \le \|A_P^* x_n\|_H \|y\|_H + C\|x_n\|_H \|y\|_H \le C\|x_n\|_{D(A_P^*)} \|y\|_H; \; y \in H.$$

where we have used (3.4). Thus

(3.13) $\quad \|A P^{-1} x_n\|_H \le C\|x_n\|_{D(A_P^*)}$

Since $A P^{-1}$ is closed on H by (3.13) and by weak closedness of $A P^{-1}$ (see [K.1]) we infer

(3.14) $\quad A P^{-1} x_n \to A P^{-1}x$ weakly in H for $x \in D\,(A_P^*)$.

Hence $A P^{-1}x \in H$ and by the Closed Graph Theorem

$$P^{-1} \in \mathcal{L}(\mathcal{D}(A_P^*); D(A)) \;\blacksquare$$

Proposition 3.3

$$B_2^* \in \mathcal{L}(D(A_P^*); U_2)$$

proof:

Setting $x = y \in D$ in (DRE) yields

$$\|B_2^* x\|_{U_2}^2 \leq \|A P^{-1} x\|_H \|x\|_H + \|P^{-1} x\|_H^2$$

by Proposition 3.2 and by 3.4

$$\leq C\,[\|x\|_{D(A_P^*)} \|x\|_H + \|x\|_H^2], \quad x \in D .$$

Now the above inequality can be extended by density (recall Proposition 3.1) to all $D(A_P^*)$. This completes the proof of the Proposition 3.3.

Proof of Lemma 3.1. The result of Lemma 3.1 follows now from (3.6) in Proposition 3.1 combined with the results of Propositions 3.2 and Proposition 3.3.

Now, we are in a position to verify the assumptions (A.2) and (A.3) of Main Theorem.

Verification of (A.2)

From Proposition 3.3 we obtain

$$B_2^* (A_P^*)^{-1} \in \mathcal{L}(H; U_2)$$

Hence by the duality

(3.15) $\quad A_P^{-1} B_2 \in \mathcal{L}(U_2; H)$

and (A.2) follows after noticing that with F given by (3.2)

(3.16) $\quad A + B_2 F = A - B_2 B_2^* P \equiv A_P .$

Verification of (A.3)

In view of (H.1) (ii) and of (3.16) it is enough to prove

Lemma 3.2

$$B_2^* e^{A_P^*(\cdot)} \in \mathcal{L}(H; L_2(0,T; U_2)) .$$

proof of Lemma 3.2

Notice first that by the result of Proposition 3.3 $B_2^* e^{A_P^* t}\colon H \to L_2((0,T); U_2)$ is densely defined (as $D(A_P^*) \subset D(B_2^* e^{A_P^*(\cdot)})$.

Next, we shall prove that this operator is closed. Indeed, let us introduce the operator $J : L_2[0, T; U_2] \to H$ given by

$$(3.17)\quad Ju \equiv \int_0^T e^{A_P\tau} B_2\, u(\tau)\, d\tau .$$

By (3.15) and by standard results (see [K.1]) J is closed. Moreover, J is also densely defined. To see this, we take $u \in H^1\,[0,T;\,U_2] \subset C\,[0T;\,U_2]$ and we compute

$$(3.18)\quad J(u) = \int_0^T \frac{d}{dt} e^{A_P\tau} A_P^{-1} B_2\, u(\tau)\, d\tau = e^{A_P T} A_P^{-1} B_2\, u(T) - A_P^{-1} B_2\, u(0) - \int_0^T e^{A_Pt} A_P^{-1} B_2\, \dot{u}\,(\tau)\, d\tau .$$

All the terms on the RHS of (3.18) are bounded in H with $u \in H^1[0T;\,U]$. Thus $H^1\,[0T;\,U_2] \subset D(J)$ and the density of D(J) follows from the density of $H^1[0T;\,U_2]$ in $L_2[0T;\,U_2]$. Hence the operator J defined by (3.17) is closed and densely defined. On the other hand it is immediate to verify that $J^* = B_2^* e^{A_P^*(\cdot)}$. Hence $B_2^* e^{A_P^*(\cdot)}: H \to L_2\,(0T;\,U_2))$ is closed and densely (on $D\,(A_P^*)$) defined, as desired.

To complete the proof of the Lemma it is enough to establish the following inequality

$$(3.19)\quad \int_0^T |B_2^* e^{A_P^*t} x|_U^2\, dt \le C\, |x|_H^2\, ; \qquad x \in D(A_P^*) .$$

Indeed, the assessment of the Lemma will follow from (3.19) and standard density argument (closedness of $B_2^* e^{A_P^*t}$ has been asserted above).

<u>proof of (3.19)</u>

By the result of Lemma 3.1 we are in a position to apply (DRE) with $x = y \in D\,(A_P^*)$. This yields

$$(3.20)\quad 2\,\text{Re}\,(A\,P^{-1}x,\, x)_H + |P^{-1}\,x|_H^2 = |B_2^*\,x|_{U_2}^2\, ; \quad x \in D\,(A_P^*)$$

Consider next

$$w(t) \equiv e^{A_P^*t}\, x \;\text{ with }\; x \in D(A_P^*)\, ,$$

so $w_t \in H$ and we have

$$(3.21)\quad \begin{cases} w_t = A_P^* w = (A^* - P\,B_2\,B_2^*)\,w \\ w(0) = x\, . \end{cases}$$

We multiply both sides of equation (3.21) by $P^{-1}w$, which by the virtue of Proposition 3.2, is in D(A).

$$(w_t,\, P^{-1}w)_H = ((A^* - P\,B_2\,B_2^*)w,\, P^{-1}w) = (w,\, A\,P^{-1}w)_H - |B_2^*w|_{U_2}^2 =$$

(integrating from 0 to T)

$$(w(T), P^{-1} w(T))_H - (x, P^{-1}x)_H = 2\,\mathrm{Re}\int_0^T (w(t), A P^{-1} w(t))_H\, dt - 2\int_0^T |B_2^* w(t)|^2_{U_2}\, dt\,.$$

Applying (3.20) with $x = w(t) \subset D(A_P^*)$ gives

$$(w(T), P^{-1}w(T))_H - (x, P^{-1}x)_H = -|P^{-1}x|^2_H - \int_0^T |B_2^* w(t)|^2_{U_2}\, dt$$

and by (3.4)

$$\int_0^T |B_2^* w(t)|^2_{U_2} \le C\,[\,|x|^2_H + |w(T)|^2_H] \le C\,|x|^2_H$$

which implies (3.19). ■

To conclude, we have verified all the assumptions of Theorem (3.1) and consequently the conclusion of Theorem 3.1 is applicable. This proves the result claimed in Main Theorem. ■

4. Applications

Kirchhoff plate with boundary feedback and boundary perturbations.

We consider on any smooth bounded $\Omega \subset R^2$,

$$(4.1)\quad \begin{cases} w_{tt} + \Delta^2 w - \rho\,\Delta w_{tt} = 0 \text{ in } (0,\infty)\times\Omega \equiv Q \\ w(0,\cdot) = w_0;\ w_t(0,\cdot) = w_1 \text{ in}\Omega \\ w|_\Sigma = 0 \\ \Delta w|_\Sigma = F(w(t), w_t(t)) + G(w(t), w_t(t)) \end{cases}$$

with $\rho > 0$ and with boundary feedback F.

Here G: $L_2(\Omega) \times L_2(\Omega) \to L_2(\Gamma)$ is Nemycki's operator associated with a scalar function g, i.e.:

(4.2) $G(y_1, y_2)(x) \equiv g(y_1(x), y_2(x))\,;\ x \in (\Gamma)$

where $g \in C^1(R \times R)$ and it satisfies the following requirements

$$(4.3)\quad \begin{cases} \text{(i)}\ \ g(0) = 0;\ g'(0) = 0\,; \\ \text{(ii)}\ \ g \text{ is of a polynomial growth in the second variable.} \end{cases}$$

The main goal of this subsection is to show that the results of Main Theorem are applicable to the present context. We shall prove that the feedback operator F: $L_2(\Omega) \times L_2(\Omega)$ given by

$$(4.4)\quad F(w, w_t) = \frac{\partial}{\partial\eta}\,\Delta P_2 \begin{bmatrix} w \\ w_t \end{bmatrix}\Big|_\Gamma$$

where the operator $P \equiv (P_1, P_2)$ solves the appropriate Riccati Equation, provides for the system (4.1), the sought after local exponential stability subject to an arbitrary perturbation g.

To accomplish our goal we need to put problem (4.1) into the abstract model (1.4).

We introduce the positive self-adjoint operator

$$\mathcal{A} h = \Delta^2 h; \quad \mathcal{D}(\mathcal{A}) = \{h \in H^4(\Omega);\ h|_\Gamma = \Delta h|_\Gamma = 0\}$$

and define the operators; $\hat{A} \equiv (I + \rho\, \mathcal{A}^{1/2})^{-1} \mathcal{A}$

$$(4.5) \quad A = \begin{vmatrix} 0 & I \\ -\hat{\mathcal{A}} & 0 \end{vmatrix}; \quad Bu = \begin{bmatrix} 0 \\ \hat{\mathcal{A}} Du \end{bmatrix}$$

where $D\colon L_2(\Gamma) \to L_2(\Omega)$ is the appropriate Green map

$$y = Dv \Leftrightarrow \{\Delta^2 y = 0 \text{ in } \Omega;\ y|_\Gamma = 0;\ \Delta y|_\Gamma = v\}.$$

Now we set

$$H \equiv [H^2(\Omega) \cap H_0^1(\Omega)] \times H_0^1(\Omega);\ U_1 = U_2 = L_2(\Gamma);$$

$$B_1 = B_2 \equiv B$$

$G\colon H^2(\Omega) \times H^1(\Omega) \to L_2(\Gamma)$ is given by (4.2).

We shall first verify the assumptions (1.1), (H-1) and (H-2).

<u>Assumption (1.1)</u>

Assumption (1.1) is equivalent showing

$$(4.6) \quad A^{-1} B \in \mathcal{L}(U, H).$$

From (4.5) we readily obtain

$$A^{-1} Bu = \begin{vmatrix} 0 & -\hat{\mathcal{A}}^{-1} \\ -I & 0 \end{vmatrix} \begin{vmatrix} 0 \\ \hat{\mathcal{A}} Du \end{vmatrix} = \begin{bmatrix} Du \\ 0 \end{bmatrix}$$

and (4.6) holds true by the elliptic regularity:

$$D \in \mathcal{L}(L_2(\Gamma); H^{5/2}(\Omega) \cap H_0^1(\Omega)) \in \mathcal{L}(U; H).$$

<u>Assumption (H-1)</u>

Part (ii) holds as $B_1 = B_2 \equiv B$. As for part (i) one can show (see [F-L-T] Appendix C)

$$(4.7)\quad B^* e^{A^* t} \begin{bmatrix} x_1 \\ x_2 \end{bmatrix} = \frac{\partial \Delta \Phi(t)}{\partial \eta} \Big|_\Gamma$$

where $\Phi(t)$ solves the corresponding homogeneous problem:

$$(4.8)\quad \begin{cases} \phi_{tt} + \Delta^2 \phi - \rho \, \Delta \, \phi_{tt} = 0 \\ \phi(0) = \phi_0 = (I + \rho \, \mathcal{A}^{1/2})^{-1} x_2 \, ; \ \phi_1 = -(I + \rho \, \mathcal{A}^{1/2})^{-1} \mathcal{A}^{1/2} x_1 \\ \phi|_\Sigma \equiv \frac{\partial \phi}{\partial \eta} \Big|_\Sigma = 0. \end{cases}$$

Thus, by (4.8), an equivalent formulation of the assumption (H-1) (i) is the inequality

$$(4.9)\quad \int_\Sigma |\Delta \phi|^2 \, d\Sigma \le C_T \, [\| \phi_0 \|^2_{H^3(\Omega)} + \| \phi_1 \|^2_{H^2(\Omega)}] .$$

It should be noted that inequality (4.9) does not follow from a priori regularity of the solution ϕ. It is an independent regularity result which holds indeed true see ([L-L], [L-T.3]) for any general smooth Ω. Thus assumption (H-1) holds true for the problem (4.1).

Assumption (H-2): By Sobolev's Imbeddings: $H^2(\Omega) \subset C(\bar{\Omega})$ and $H^{1/2}(\Gamma) \subset L^p(\Gamma)$ for any $p > 0$. Since the trace operator $y|_\Gamma$ is bounded from $H^1(\Omega)$ into $H^{1/2}(\Gamma)$, the operator G given by (4.2) is continuously Frechet differentiable from $H^2(\Omega) \times H^1(\Omega) \to L_2(\Gamma)$. ■

We verify next that the assumptions (1.8) and (1.9), leading to the solvability of Riccati Equation, are satisfied as well.

Assumption (1.8): By the result of [L-T.3], the exact controllability of problem (4.1) holds true for any $T > 0$ on the space $H = [H^2(\Omega) \cap H^1_0(\Omega)] \times H^1_0(\Omega)$.

Assumption (1.9) is satisfied automatically since A is a group.

Thus all the assumptions of Main Theorem have been verified and we are in a position to state the main result.

Theorem 4.1

Consider the system (4.1) with the feedback operator F: $H^2(\Omega) \times H^1(\Omega) \to L_2(\Gamma)$ given by

$$F(w, w_t) \equiv - \frac{\partial}{\partial \eta} \Delta P_2 (w, w_t) |_\Gamma$$

where $P \equiv (P_1, P_2)$ is the solution to (ARE) with A and B as above. Then for any perturbations G subject to (4.3), system (4.1) is locally exponentially stable in the topology of $H^2(\Omega) \times H^1(\Omega)$.

Remark

Other dynamics, like wave equations or plate equations with uncontrolled nonlinear perturbations on the boundary can be treated in a similar manner. Indeed, one can verify that for these systems (see [F-L-T]) all abstract assumptions (H-1), (H-2), (A-1) - (A-3) are satisfied. Therefore, the conclusion of Main Theorem is applicable to yield the sought after stability results with Riccati feedback applied on the boundary.

References:

[B.1] A. V. Balakrishnan, *Applied Functional Analysis*, Springer-Verlag, 1981.

[F.1] F. Flandoli, "Riccati Equatins Arising in a Boundary Control Problem with Distributed Parameters", *SIAM J. Contr. Opt.* 22 (1984), 76-86.

[F-L-T] F. Flandoli, I. Lasiecka, R. Triggiani, "Algebraic Riccati equations with non-smoothing observation arising in hyperbolic and Euler-Bernoulli equations," *Ann. Matem. Pura e Appl.* Vol. CLiii (1988), 307-382.

[K.1] T. Kato, *Perturbation Theory for Linear Operators*, Springer-Verlag (1966), Berlin.

[L.1] I. Lasiecka, "Boundary stabilization of hyperbolic and parabolic equations with non-linearly perturbed boundary conditions," *J. Diff. Eq.*, Vol. 75 No. 1, (1988), 53-871.

[L-L] I. Lagnese, J. L. Lions, *Modelling Analysis and Control of Thin Plates*, Masson, 1988.

[L-T.1] I. Lasiecka, R. Triggiani, "Finite rank, relatively bounded perturbations of Co-semigroups. Part I: well posedness and boundary feedback hyperbolic dynamics," *Annali Scuola Normale di Pisa*, Vol Xii, No. 4, (1985).

[L-T.2] I. Lasiecka, R. Triggiani, "Dirichlet boundary control problem for parabolic equations with quadratic cost: Analyticity and Riccati's feedback synthesis," *SIAM J. Contr. Optimiz.* 21, (1988), 41-68.

[L-T.3] I. Lasiecka, R. Triggiani, "Regularity, exact controllability and uniform stabilization of Kirchhoff plates via only the bending moment," Proceedings of 28 IEEE - CDC Conference, Tampa 1989.

[T.1] R. Triggiani, "A^ε - bounded, finite rank perturbations of s.c. group generators: counterexamples to generation and to another condition for well-posedness," Lectures Notes in Mathematics #1076, Springer-Verlag 1984, pp. 227-233; Proceedings of Workshop on Operator semigroups and applications, Retzholf, Austria, June 1983.

SOME TWO-DIMENSIONAL BOUNDARY SHAPE OPTIMIZATION PROBLEMS FOR DISTRIBUTED PARAMETER SYSTEMS

A.Y. Mednikov, V.A. Troitsky
Polytechnical Institute
Polytechnicheskaya, 29, Leningrad, USSR.

There are many investigations about boundary-shape optimization problems (BSOP) and many analytical and numerical methods to solve them (Banichuk, 1970). This report deals with only four two-dimensional boundary-shape optimization problems for elastic bodies: namely, the boundary-shape optimization problems for oscillating membrane, vibrating thin plates, and bending thin plates and torsional prismatic bars. It is well known that many BSOP have multi-extremal solutions. For this reason, we attempt to obtain and analyze the second variation. At the same time we discuss numerical results which were derived by gradient algorithms. Also the FEM and BEM algorithms for solving corresponding problems are discussed.

Let us denote the boundary to be optimized by Γ, the inner domain bounded with contour Γ by Ω and the outer unit normal vector by n as illustrated in figure 1.

Figure 1:

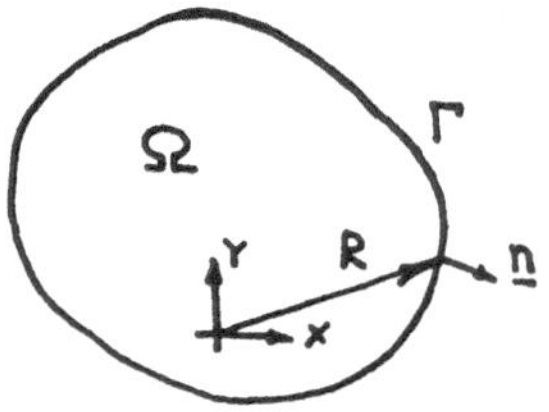

Therefore we can write the variation of a double integral with variable boundary as

$$\delta\iint_{\Omega} F(R)d\Omega = \iint_{\Omega}\left[\delta F + \nabla F\cdot\delta R\right]d\Omega + \iint_{\Omega} F\ \delta(d\Omega) =$$

$$= \iint_{\Omega}\left[\delta F + \nabla F\cdot\delta R\right]d\Omega + \iint_{\Omega} F\ \nabla\cdot\delta R\ d\Omega =$$

$$=\iint_{\Omega}\delta F d\Omega + \int_{\Gamma} F\ n\cdot\delta R\ d\Omega = \iint_{\Omega}\delta F d\Omega + \int_{\Gamma} F\ \delta n\ d\Omega.$$

In these formulae the operator ∇ is defined as $\nabla = i\partial/\partial x + j\partial/\partial y$, also we have denoted $n\cdot\delta R=\delta n$, where δR is the variation of the boundary Γ.

Shape optimization problem for oscillating membrane. The optimization problem is to find the boundary contour that minimizes the first natural frequency of membrane. The area of membrane is fixed (and equals to S^*). Therefore

$$S = \iint_{\Omega} d\Omega = S^*.$$

If we use Reyleigh formula then we can write functional to minimize as

$$\Pi = \frac{1}{2}\iint_{\Omega}(\nabla w)^2 d\Omega.$$

We suppose the isoperimetric constraint

$$T = \frac{1}{2}\iint_{\Omega} w^2 d\Omega = 1$$

to take place. Here Π and T are amplitude values of potential and kinetic energies, $w(x,y)$ – displacement of membrane points.

We are searching for the boundary contour of membrane that corresponds to minimum of the first natural frequency.

The adjoint functional of this optimality problem is

$$\mathfrak{J} = \frac{1}{2}\iint_{\Omega}(\nabla w)^2 d\Omega + \beta\left[\frac{1}{2}\iint_{\Omega} w^2 d\Omega - 1\right] + \gamma\left[\iint_{\Omega} d\Omega - S^*\right]$$

where β and γ are the constant Lagrange multipliers. The first variation of this functional can be transformed to

$$\delta\mathfrak{J} = -\iint_{\Omega}\left[\nabla^2 w - \beta w\right]\delta w d\Omega + \int_{\Gamma_0}\left[\gamma - \frac{1}{2}\left(\frac{\partial w}{\partial n}\right)^2\right]\delta n\ d\Gamma.$$

We have used the condition $w=0$ on contour Γ.

In accordance with the first necessary condition of the calculus of variations (rule of multipliers) this first variation must be equal to zero. Therefore $\delta\mathfrak{J}=0$.

If we carry out usual variational transformations we obtain the equation

$$\nabla^2 w - \beta w = 0 \quad \text{in } \Omega,$$

$$\gamma - \frac{1}{2}\left(\frac{\partial w}{\partial n}\right)^2 = 0 \text{ on } \Gamma.$$

As can be easily shown the circular boundary contour satisfies this equation.

The second variation of functional at the point of extremum can be transformed to

$$\delta^2 \mathfrak{J} = \iint_\Omega \left[(\nabla \delta w)^2 + \beta (\delta w)^2 \right] d\Omega.$$

This formula together with Reyleigh formula show that

$$\delta^2 \mathfrak{J} \geq 0.$$

Therefore circular membrane has minimal first natural frequency of all the membranes of the same area.

We consider two problems from a variety of boundary optimization problems for thin plates namely contour optimization for free transverse vibrations, and boundary shape optimization for bending of a clamped plate.

Contour optimization for free transverse vibrations. The functional in this problem is the first natural frequency. If we use the Rayleigh formula again we obtain the functional

$$\Pi = \frac{D}{2} \iint_\Omega \left[(1-\nu) \nabla\nabla w \cdot\cdot \nabla\nabla w + \nu (\Delta w)^2 \right] d\Omega$$

and isoperimetric constraints

$$T = \frac{\rho}{2} \iint_\Omega w^2 d\Omega = 1,$$

$$S = \iint_\Omega d\Omega = S^*.$$

We again assume that the area of the plate is fixed. In these formulas $w(x,y)$ is the displacement of the middle surface points, ν – the Poisson ratio, ρ – the density of material, E – the Young modulus, and $D = Eh^3/12(1-\nu^2)$ is the rigidity of the plate.

The adjoint functional of this optimality problem is

$$\mathfrak{J} = \frac{D}{2} \iint_\Omega \left[(1-\nu) \nabla\nabla w \cdot\cdot \nabla\nabla w + \nu (\Delta w)^2 \right] d\Omega + \beta \left[\iint_\Omega w^2 d\Omega - 1 \right] + \gamma \left[\iint_\Omega d\Omega - S^* \right]$$

where β and γ are the constant Lagrange multipliers. The first variation of this functional can be transformed to $\delta \mathfrak{J} = -D \iint_\Omega \left[\Delta\Delta w + \frac{\beta}{D} w \right] \delta w \, d\Omega +$

$$+D\int_{\Gamma_o}\Big[-n\cdot\nabla\Delta w\delta w + n\cdot((1-\nu)\nabla\nabla w + \nu\Delta w E)\cdot\nabla\delta w +$$

$$+ \frac{1}{2}((1-\nu)(\nabla\nabla w\cdot\cdot\nabla\nabla w)+\nu(\Delta w)^2)\delta n + \frac{1}{D}(\frac{1}{2}\beta w^2+\gamma)\delta n\Big]d\Gamma,$$

where E is the plane unit tensor.

Here the first necessary condition of the calculus of variations $\delta\mathfrak{J}=0$ leads to equations

$$R = \Delta\Delta w + \frac{\beta}{D}w = 0 \quad \text{in } \Omega,$$

$$T_p=0 \quad \text{on } \Gamma.$$

We consider the different boundary conditions now. For clamped boundary contour we have equalities

$$w = 0\,, \quad \frac{\partial w}{\partial n} = 0 \quad \text{on } \Gamma$$

consequently

$$\delta_* w = \delta w + \frac{\partial w}{\partial n}\delta n = 0\,, \quad \delta_*\frac{\partial w}{\partial n} = \delta\frac{\partial w}{\partial n} + \frac{\partial^2 w}{\partial n^2}\delta n = 0\,.$$

Here the asterisk marks the full variation. Therefore we have

$$\delta w = -\frac{\partial w}{\partial n}\delta n = 0 \text{ and } \delta\frac{\partial w}{\partial n} = -\frac{\partial^2 w}{\partial n^2}\delta n \text{on } \Gamma$$

and from equality $T_p=0$ on Γ we obtain

$$-\frac{\partial^2 w}{\partial n^2} = \frac{\gamma}{D} \quad \text{on } \Gamma.$$

It can be shown that the circular plate is optimal. The second variation in this case can be written as

$$\delta^2 J = D\Big\{\iint_\Omega\Big[(\Delta\delta w)^2 + \frac{\beta}{D}(\delta w)^2\Big]d\Omega - \int_\Gamma \frac{\partial^2 w}{\partial n^2}\Big[\frac{\partial^2 w}{\partial n^2} + 2\frac{\partial^2 w}{\partial n^2}k\Big]\delta n^2 d\Gamma\,.$$

Boundary shape optimization for bending of a clamped plate. We shall consider the second shape optimization problem of a thin plate as to minimize the functional

$$E(\Omega) = \iint_D (W(\Omega)-W_d)^2 do,$$

where $W_d \in L^2$ is the given function and the domains D and Ω ($D\subset\Omega$) are illustrated on figure 2.

Figure 2:

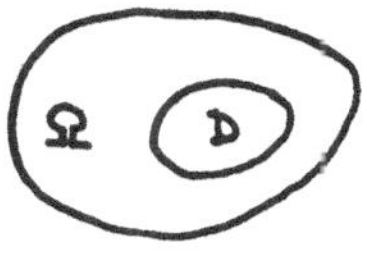

We again assume that the area of the plate is fixed, i.e.

$$S = \iint_{\Omega} d\Omega = S^{*}$$

and also we consider that the displacement of the middle surface points $W(\Omega)$ satisfies the equalities

$$\Delta^2 W = f \quad \text{in } \Omega ,$$

$$W = 0 \ , \ \frac{\partial W}{\partial n} = 0 \quad \text{on } \Gamma .$$

The adjoint functional of this optimality problem is

$$\mathfrak{J} = \iint_{D} (W(\Omega)-W_d)^2 do + \iint_{\Omega} p(\Delta^2 W - f) d\Omega + \beta\Big(\iint_{\Omega} d\Omega - S^{*}\Big) ,$$

where the function $p \in H_0^2(\Omega)$ is the Lagrange multiplier. The first variation of this functional can be transformed to

$$\delta\mathfrak{J} = 2\iint_{D} \delta W(W(\Omega)-W_d) do + \iint_{\Omega} \Delta^2 p \ \delta W d\Omega +$$

$$+ \int_{\Gamma} \Big[\Delta p \frac{\partial \delta W}{\partial n} - \frac{\partial \Delta p}{\partial n} \delta W + \beta\Big] \delta n d\Gamma.$$

Taking into account the boundary condition on a clamped edge we can write the variations on contour Γ as $\delta W = - \frac{\partial W}{\partial n}\delta n, \quad \frac{\partial \delta W}{\partial n} = - \frac{\partial^2 W}{\partial n^2}$

Therefore the first variation of the adjoint functional may be obtained as

$$\delta\mathfrak{J} = 2\iint_{\Omega} (\Delta^2 p + 2\chi_d(W(\Omega)-W_d))\delta W d\Omega - \int_{\Gamma} \Big[\Delta p \frac{\partial^2 W}{\partial n^2} - \frac{\partial \Delta p}{\partial n^2} \frac{\partial W}{\partial n} + \beta\Big] \delta n d\Gamma,$$

where χ_d is the characteristic function.

Once again, the first necessary condition of the calculus of variations $\delta\mathfrak{J}=0$ leads to equation

$$\Delta p \frac{\partial^2 W}{\partial n^2} - \frac{\partial \Delta p}{\partial n} \frac{\partial W}{\partial n} + \beta = 0 \quad \text{on } \Gamma,$$

where $p \in H_0^2(\Omega)$ is the solution of the adjoint problem

$$\Delta^2 p + 2\chi_d(W(\Omega)-W_d) \qquad \text{in } \Omega ,$$

$$p = 0 \ , \ \frac{\partial p}{\partial n} = 0 \qquad \text{on } \Gamma.$$

The second variation in the shape optimization problem that is discussed can be written as

$$\delta^2\mathfrak{J} = 2\iint_{\Omega} \chi_d \delta W^2 d\Omega + \int_{\Gamma} \left[\frac{\partial^2 p}{\partial n^2}\frac{\partial^3 W}{\partial n^3} - \frac{\partial^3 p}{\partial n^3}\frac{\partial^2 W}{\partial n^2} + k\frac{\partial^2 p}{\partial n^2}\frac{\partial^2 W}{\partial n^2}\right]\delta n^2 d\Gamma,$$

where k is the curvature of the boundary. To obtain the last formulae we implied the usual double integral transformations and the property of variations δw that is $\Delta^2\delta W = 0$.

Numerical solution of the optimization problem discussed is based on gradient algorithm of order one. The value of the gradient of minimizing functional is represented as

$$\delta\mathfrak{J} = -\int_{\Gamma} \left[\Delta p \frac{\partial^2 W}{\partial n^2} + \beta\right]\delta n d\Gamma,$$

where $W \in H_0^2(\Omega)$ and $p \in H_0^2(\Omega)$ are functions always introduced. Therefore at each point we have to transform the designing boundary in the direction of outer normal with respect to the value

$$\delta n \simeq \frac{\partial^2 p}{\partial n^2}\frac{\partial^2 W}{\partial n^2} + \beta, \qquad \text{where } \beta = -\int_{\Gamma} \frac{\partial^2 p}{\partial n^2}\frac{\partial^2 W}{\partial n^2} d\Gamma \Big/ \int_{\Gamma} d\Gamma.$$

Some numerical results are pictured on figure 3. To simplify the problem the D-region was considered as a point.

Figure 3:

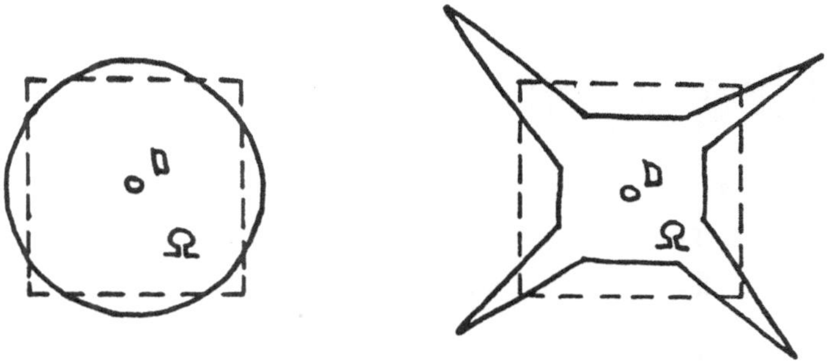

The left figure presents the case when $\inf_{\Omega_i} \{\mathfrak{J}_i\} > 0$, i.e. the lower limit of minimizing functional is larger than zero (absolute minimum). In this case there is only one optimal solution that is the circular boundary. The right figure represents the case when $\inf_{\Omega_i} \{\mathfrak{J}_i\} = 0$, i.e. the lower limit of minimizing functional achieves the absolute minimum value that is equal to zero. In this case the optimality problem has no one solution and each of them is optimal. There the extra mass is in some way distributed into outer space.

The model was discretized with only boundary elements. The analysis (which was made by the authors) of BEM- and FEM- algorithms due to plate bending problem shows that BEM is more efficient when the edge of plate is clamped.

Torsion problem. Cross section of prismatic bar is shown in figure 4.

Figure 4:

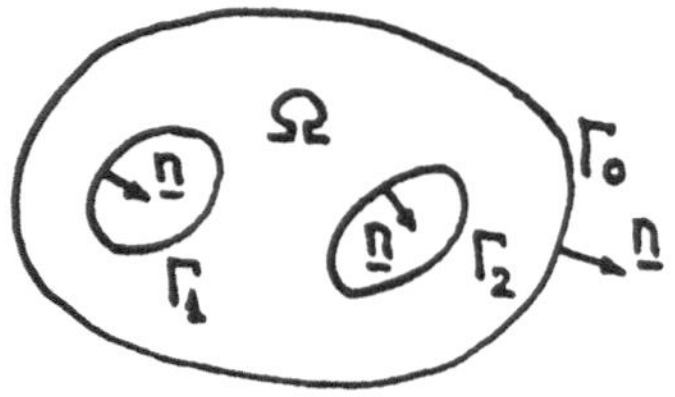

$\Gamma = \cup \Gamma_k, \ k=\overline{0,n}$

Contours Γ_k, $k=\overline{1,n}$ are fixed. Outer contour Γ_0 may be changed. Mass of bar is fixed. Therefore

$$S = \iint_{\Omega} d\Omega = S^*.$$

Here S is cross section area and S^* - given value.

If we maximize geometrical torsion stiffness of bar, the functional can be written as

$$\mathfrak{C} = 2\left[\iint_{\Omega} \Phi d\Omega + \sum_{k=1}^{n} C_k S_k \right]$$

where $\Phi=\Phi(x,y)$ is stress function of the torsion problem, S_k- area bounded by contour Γ_k, C_k - constant value such that

$$\left[\Phi\right]_{\Gamma_k} = C_k, \ k=\overline{1,n}; \qquad \left[\Phi\right]_{\Gamma_0} = 0.$$

For stress function Φ we have equality

$$\tau = \mu\alpha\underline{\nabla}\Phi \times \underline{k}$$

Here $\tau=(\tau_x,\tau_y)$ is tangential stresses, α - torsional angle, ∇ - two-dimensional operator "nabla", k - unit vector in z direction, μ - shear modulus.

The problem is expressed as finding the shape of outer boundary contour Γ_0 to maximize geometrical rigidity, subject to constraint of fixed area.

Real stress state of bar gives minimum of additional work Ψ, that is given by

$$\Psi = \frac{1}{2}\iint_{\Omega}\left[(\nabla\Phi)^2 - 4\Phi\right]d\Omega - 2\sum_{k=1}^{n} C_k S_k.$$

Geometrical rigidity is connected with the additional work by the equality

$$\mathfrak{C} = -2\min_{\Phi} \Psi.$$

Using the variation calculus we have to compose the adjoint functional

$$\mathfrak{J} = \min \Psi + \beta\left[\iint_{\Omega} d\Omega - S^*\right]$$

Here β is Lagrange multiplier. The first variation of these functional may be obtained as

$$\delta\mathfrak{J} = + \int_{\Gamma_o}\left[\frac{1}{2}(\nabla^o\Phi)^2 + \beta\right]\delta n d\Gamma = - \iint_{\Omega}\left[\Delta^o\Phi + 2\right]\delta\Phi d\Omega + \int_{\Gamma_o}\frac{\partial\Phi}{\partial n}\,\delta\Phi d\Gamma +$$

$$+ \sum_{k=1}^{n}\left[\int_{\Gamma_k}\frac{\partial\Phi}{\partial n}\,\delta C_k d\Gamma - 2\sum_{k=1}^{n}\delta C_k S_k + \int_{\Gamma_o}\left[\frac{1}{2}(\nabla^o\Phi)^2 + \beta\right]\delta n d\Gamma\right.$$

The formulas

$$\delta\Phi = -\frac{\partial\Phi}{\partial n}\,\delta n\ , \qquad (\nabla^o\Phi)^2 = \left[\frac{\partial\Phi}{\partial n}\right]^2 \quad \text{on } \Gamma_o,$$

imply the equality

$$\delta\mathfrak{J} = -\iint_{\Omega}\left[\Delta^o\Phi+2\right]\delta\Phi d\Omega + \sum_{k=1}^{n}\int_{\Gamma_k}\left[\frac{\partial\Phi}{\partial n} - 2S_k\right]\delta C_k d\Gamma - \int_{\Gamma_o}\left[\frac{1}{2}\left[\frac{\partial\Phi}{\partial n}\right]^2 - \beta\right]\delta n d\Gamma.$$

In accordance with the first necessary condition of functional minimum (multipliers rule) this first variation must be equal to zero. Therefore $\delta\mathfrak{J}=0$.

Usual transformations gives us Euler equation

$$\Delta^o\Phi = -2 \qquad \text{in } \Omega,$$

conditions

$$\int_{\Gamma_k}\frac{\partial\Phi}{\partial n_k}d\Gamma = 2S_k\ ,\ k=\overline{1,n},$$

and equality

$$\frac{1}{2}\left[\frac{\partial\Phi}{\partial n}\right]^2 - \beta = 0 \qquad \text{on } \Gamma_0.$$

This equality determines optimal shape of contour Γ_0.

These results are well known. Our way to achieve them is simpler.

If we have simply connected cross section then the last equality is satisfied on circular boundary. The second variation of our functional can be written as

$$\delta^2 \mathfrak{J} = 2 \iint_{\Omega} \left[(\nabla \times k) \delta\Phi \right]^2 d\Omega.$$

It is nonnegative. Therefore the geometric rigidity of circular section is maximal.

For multiconnected section we have to use numerical method for solution of optimality problem. The gradient of minimizing functional can be written as

$$\delta \mathfrak{J} = \int_{\Gamma_0} \left[- \left(\frac{\partial \Phi}{\partial n} \right)^2 + \beta \right] \delta n d\Gamma ,$$

where function Φ was already introduced. Therefore like in above section at each point we have to transform the designing boundary in the direction of outer normal with respect to the value

$$\delta n \approx \alpha \left[\left(\frac{\partial \Phi}{\partial n} \right)^2 - \beta \right],$$

where $\alpha \in (0,1]$. Constant value of β is defined of isoperimetric constraint as

$$\delta S^* = \int_{\Gamma_0} \delta n d\Gamma = 0, \text{ therefore } \beta = -\int_{\Gamma_0} \left(\frac{\partial \Phi}{\partial n} \right)^2 d\Gamma \Big/ \int_{\Gamma_0} d\Gamma = \left(\frac{\partial \Phi}{\partial n} \right)^2_{cp.}$$

When the condition

$$\int_{\Gamma_0} |\delta n| d\Gamma \le \varepsilon$$

is satisfied the optimal problem considered to be solved.

The Φ function is obtained in a following form

$$\Phi = \Phi_0 + \sum_{k=1}^{n} C_k \Phi_k,$$

where functions Φ_i, $i=\overline{0,n}$ are the solution of the problem as

$$\Delta^0 \Phi_0 = -2 \, , \quad \Delta^0 \Phi_k = 0 \quad \text{in } \Omega \, , \quad k=\overline{1,n};$$

$$\Phi_0 = 0 \qquad \text{on } \Gamma_k \, , \quad k=\overline{0,n};$$

$$\Phi_i = \delta_{ik} \qquad \text{on } \Gamma_k \, , \quad k=\overline{1,n} \, , \quad i=\overline{1,n}.$$

The unknown constants C_k ($k=\overline{1,n}$) can be obtained from the equality

$$\int_\Gamma \left[\frac{\partial\Phi_o}{\partial n_k} + \sum_{i=1}^{n} C_i\frac{\partial\Phi_i}{\partial n_k}\right] d\Gamma = 2S_k,$$

that is transformed to the system of linear equations:

$$\sum_{k=1}^{n} B_{ki}C_i = B_k,$$

where

$$B_{ki} = \int_\Gamma \frac{\partial\Phi_i}{\partial n_k} d\Gamma\ ,\quad B_k = 2S_k - \int_\Gamma \frac{\partial\Phi_o}{\partial n_k} d\Gamma,\qquad k=\overline{1,n},\ i=\overline{1,n}.$$

We prefer the BEM algorithm because of some disadvantages of FEM: namely the FEM requirement of mesh regeneration on each optimization step is not accurate, and it computes slowly the design variables such as $\partial\Phi/\partial n$ at boundary points.

Discretization of a model was made by constant elements. Our analysis shows that it is simpler to achieve accuracy by increasing the number of constant elements rather than by implying the linear elements.

The computation of geometrical stiffness

$$\mathfrak{G} = \iint_\Omega \Phi do$$

by BEM-algorithm may be overcome in the following way. Taking into account the known formula

$$\Phi = 2\iint_\Omega \Phi^* do + \int_\Gamma \frac{\partial\Phi}{\partial n}\Phi^* d\Gamma - \sum_{k=1}^{n}\int_{\Gamma_k} \Phi_k \frac{\partial\Phi^*}{\partial n} d\Gamma \text{ in } \Omega,$$

where $\Phi^* = \frac{1}{2\pi}\, ln\frac{1}{r}$ is the fundamental solution of Laplace equation we may write

$$\mathfrak{G} = \iint_\Omega \Phi do = \iint_\Omega \left(2\iint_\Omega \Phi^* do \right) do + \iint_\Omega \left(\int_\Gamma \frac{\partial\Phi}{\partial n}\Phi^* d\Gamma \right) do.$$

Having implied the usual double integral transformations we may obtain the needed formulae as

$$\mathfrak{G} = 2\left\{\frac{1}{2\pi}\int_\Gamma \left(\int_\Gamma \left(\frac{r}{2}\left[ln\frac{1}{r}+\frac{1}{2}\right]\frac{\partial\Phi}{\partial n} + \frac{r^2}{3}\left[ln\frac{1}{r}+\frac{5}{6}\right]\cos(n_0,r)\right]\cos(n,r)d\Gamma\right] d\Gamma + \right.$$

$$\left. + \sum_{k=1}^{n} C_k S_k \right\},$$

where n_0 denotes the outer normal at current point of the external integral.

Several numerical results are illustrated in Figure 5. We observe the solution to the optimal problem depends on the initial form of the designed boundary. The results for the cross sections with one shaft are in accordance with the the same results obtained by other authors.

Figure 5:

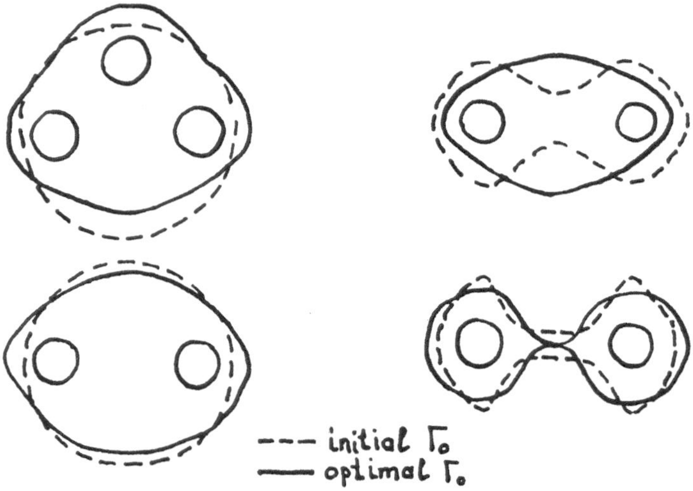

In conclusion, the analysis of the second variation applied to simple problems confirms well-known results. At the same time all attempts to obtain and to analyse the second variation in more complex problems were discharged because of their hopelessness. Applying BEM to solve BSOPs is a good idea and has only a few problems. It seems the FEM is, in general, more adequate because it is a more universal tool of numerical analysis.

References:

1. Баничук Н.В. Оптимизация формы упругих тел.- М.:Наука,1980, 302с.
2. Лурье А.И. Теория упругости.- М.:Наука, 1970, 940с.
3. Троицкий В.А. Оптимальные процессы колебаний механических систем. - Л.: Машиностроение, 1976,248с.
4. Haftka R.T., Grandhi R.V. Structural shape optimization - a survey. - Comp. Meth. in Appl. Mech. and Eng.,57(1986)91-106.
5. Soares C.A, Rodrigues H.C. Faria L.M. Optimization of the shape of solid and hollow shafts using BEM. BE 5, Proc. 5th Int. Conf. Hirosima, nov. 1983.

INVERSE PROBLEM OF DYNAMICS FOR SYSTEMS DESCRIBED BY PARABOLIC INEQUALITY

Yu.S. Osipov

Institute of Mathematics and Mechanics
of the Ural Scientific Center
Academy of Sciences of the USSR, Sverdlovsk

The considered problem is concerned with the following questions.

Let t be the time variable. Consider an evolutional system Σ on an interval $T = [t_0,\theta]$. We are interested in some unknown characteristic $\xi_1(t)$, $t \in T$ of the system (e.g., ξ_1 may be a collection of some parameters of the system, or of some disturbances acting on the system or of controls etc.). We are to reconstruct $\xi_1(t)$ on the basis of measurements of some other characteristic $\xi_2(t)$, $t \in T$ of the system Σ. The results of measurements $\varsigma(t)$ are not precise, the error being estimated by h.

The smaller h is, the more precise should be the reconstruction (in the appropriate sense). This is the stability property of the reconstruction algorithm D_h.

We consider two types of reconstruction problems. In the problems of the first type (which we call problems of program reconstruction) the measurements $\varsigma(t)$ are known for all $t \in T$ at once. Hence the input of the reconstruction algorithm is the function $\varsigma(t)$, $t_0 \le t \le \theta$. The output of D_h is a function $\xi_1^{(h)}(t)$, $t_0 \le t \le \theta$ close (in a suitable sense) to the characteristic $\xi_1(t)$, $t_0 \le t \le \theta$ for h small enough.

In problems of the second type (we call them problems of dynamical reconstruction) the characteristic ξ_1 is to be restored simultaneously with the process of system motion. Here in every current moment t the input of the algorithm D_h is the previous history $\varsigma_t = \varsigma_t(\cdot) = \{\varsigma(\tau),\ t_0 \le \tau < t\}$ of the measurements ς made prior to the moment t. The output of D_h in the moment t is a function

$$\xi_{1t}^{(h)}(\cdot) = \{\xi_1^{(h)}(\tau),\ t_0 \le \tau < t\} \ ,$$

which approximates (in the proper sense) the characteristic

$$\xi_1(\tau),\ t_0 \le \tau \le t \ , \ \text{for small } h \ .$$

Here D_h is to satisfy the property of physical realizability [2], [3]: if $\varsigma^{(1)}(\tau)$, $t_0 \le \tau \le t_1$ and

$\varsigma^{(2)}(\tau)$, $t_0 \le \tau \le t_2$ are such that

$$\varsigma_{t_*}^{(1)} = \varsigma_{t_*}^{(2)},\ t_* \le \min\ \{t_1,t_2\}\ ,$$

then the functions $D_h\varsigma_t^{(1)}(\cdot)$, $D_h\varsigma_t^{(2)}$ are equal on $[t_0,t_*)$.

Below we consider a problem of the second type for a system described by a parabolic inequality. We develop further the method for dealing with such kind of problems proposed in [1-3]. The method is based on some ideas of positional control theory [14-17] and ill-posed problems theory [18].

The present paper is connected with [1-13].

Let $\mathbf{V}$ and $\mathbf{H}$ be real Hilbert spaces, $\mathbf{V}^*$ and $\mathbf{H}^*$ be the spaces dual to $\mathbf{V}$ and $\mathbf{H}$ respectively. We identify $\mathbf{H}$ with $\mathbf{H}^*$. It is supposed that $\mathbf{V} \subset \mathbf{H}$ is dense in $\mathbf{H}$ and is embedded into $\mathbf{H}$ continuously. Denote by $(\cdot,\cdot)_{\mathbf{H}}$ and $|\cdot|_{\mathbf{H}}$ ($(\cdot,\cdot)_{\mathbf{V}}$ and $|\cdot|_{\mathbf{V}}$) the scalar product and the corresponding norm in $\mathbf{H}$ (in $\mathbf{V}$).

Let t be the time variable, $t \in T = [t_0,\theta]$. Consider on T a control system Σ. The state of the system is $y(t) \in \mathbf{V}$. The evolution of the state is given by the following conditions for almost all $t \in T$ the inequality holds ([19,20]):

$$(\dot{y}(t),\ y(t) - \omega)_{\mathbf{H}} + a(y(t),\ y(t)) + \phi(y(t)) - \phi(\omega) \le (Bu(t) + f(t),\ \omega)_{\mathbf{H}}\ \ \forall \omega \in \mathbf{V} \quad (1.1)$$

and

$$y(t_0) = y_0\ . \quad (1.2)$$

Here $a(\omega_1,\omega_2)$ is a continuous on $\mathbf{V}$ bilinear symmetrical form satisfying for some $c_1 > 0$ the condition

$$a(\omega,\omega) \ge c_1|\omega|_{\mathbf{V}}^2\ ; \quad (1.3)$$

$\phi{:}\mathbf{V} \to (-\infty,+\infty]$ is a convex proper lower semicontinuous function (or $\phi{:}\mathbf{H} \to (-\infty,+\infty]$ is a convex proper lower semicontinuous function satisfying the regularity condition [21,22]; $B : \mathbf{U} \to \mathbf{H}$ is a linear continuous operator, $\mathbf{U}$ is a uniformly convex real Banach space; $f \in L^2(T;\mathbf{H})$; $u(\cdot)$ is a control, i.e. measureable on T function for almost all $t \in T$ having values in bounded closed convex set $P \subset \mathbf{U}$; $y_0 \in \{\omega \in \mathbf{V} : \phi(\omega) < +\infty\}$. Under the above assumptions in $\mathbf{W}^{1,2}(T;\mathbf{H}) \cap L^2(T;\mathbf{V})$ there exists a unique function $y(t) = y(t;t_0,y_0,u(\cdot))$, $t \in T$, satisfying (1.1), (1.2) (see [19-22]). We call it a motion of system Σ from the initial state y_0 corresponding to control $u(\cdot)$.

Consider the following problem of dynamical reconstruction. Let $\mathbf{V} = \mathbf{H}_0^1(\Omega)$ (or $\mathbf{V} = \mathbf{H}^1(\Omega)$), $\mathbf{H} = L^2(\Omega)$,$\mathbf{U} = L^2(\Omega)$,B be the identity operator (see notation in [19,20]). Now in (1.1) we take

$$y(t) = y(t,\cdot) = \{y(t,x),\ x\in\Omega\}\ ,$$

$$\dot{y}(t) = \partial y(t,\cdot)/\partial t,\ u(t) = u(t,\cdot)\ .$$

Let the control u be of the form

$$u(t) = u(t,x) = \chi_{G(t)}(x) \times u^0(t,x) \tag{1.4}$$

Here $G(t) \subset \Omega$ is such that the set $\{(t,x) : t\in T,\ x\in G(t)\}$ is Lebesgue measureable; χ_G is the characteristic function of G; the function u^0 satisfies the inequality

$$0 < \beta_1 \le u^0(t,x) \le \beta_2,\ t\in T,\ x\in\Omega\ , \tag{1.5}$$

where β_1, β_2 are positive numbers.

Let the measurement of the system state $y_*(t) = y_*(t,\cdot)$ be possible in every current moment t, the measurement result $\varsigma(t) = \varsigma(t,\cdot)$ satisfying the estimation

$$|\varsigma(t,\cdot) - y_*(t,\cdot)|_{L^2(\Omega)} \le h\ . \tag{1.6}$$

Suppose that the motion being observed is generated by the unique control of the type (1.4), (1.5)

$$u_*(t,x) = \chi_{G_*(t)} u_*^0(t,x),\ t\in T,\ x\in\Omega\ .$$

Consider the problem of dynamical reconstruction with

$$\xi_1(t) = \{u_*(t)\ ;\ S_*(t)\}\ ,$$

$$S_*(t) = \{(\tau,x) : \tau\in[t_0,t),\ x\in G_*(\tau)\}\ ;$$

$$\xi_2(t) = y(t,\cdot)\ .$$

Remark 1.1. Let e.g., (1.1), (1.2) describe the process of diffusion of a substance in a domain Ω and $y(t,\cdot)$ be the concentration of substance in Ω in the moment t. Then we deal with the reconstruction of intensity of the substance sources and their location (see [12]).

We proceed the following way (see [12, 13]). To the system Σ we put into correspondence a control system Σ_1 (the model) which is a copy of Σ.

$$(z(t),\ z(t) - \omega)_{L^2(\Omega)} + a(z(t),\ z(t)) \tag{1.7}$$

$$- \omega) + \phi(z(t)) - \phi(\omega) \le (v(t) + f(t),\omega)_{L^2(\Omega)} \quad \forall w \in V$$

$$z(t_0) = y_0 \, .$$

The control $v(\cdot) \in L^2(T\,;\,L^2(\Omega))$ in the model is chosen for almost all $t \in T$ from convex bounded closed set P which contains all the $L^2(\Omega)$ functions of the form $\chi_B \cdot g(x)$ where $B \subset \Omega$ is a measurable set, $g(\cdot)$ is a measurable function, $g : \Omega \to [\beta_1,\beta_2]$.

Consider a partition τ_i of interval T,

$$t_0 = \tau_0 < \tau_1 < \cdots < \tau_m = \theta \, ;$$

$$m = m(h),\ \delta(h) = \max_i(\tau_{i+1} - \tau_i),\ \delta(h) \le ch,\ c = const > 0 \, .$$

Take

$$v(t) = v^{(h)}(t) = v_i,\ \tau_i \le t < \tau_{i+1}\,, \quad i = 1,\ldots,m$$

where v_i are (the unique) points of minimum of the functional

$$\psi(p) = 2(z(\tau_i\,;\,t_0,\,y_0,\,v(\cdot)) - \varsigma(\tau_i),\,p)_{L^2(\Omega)} + \alpha(h)|p|^2_{L^2(\Omega)} \, .$$

The function $\alpha(h) > 0$; $\alpha(h) \to 0$, $h/\alpha(h) \to 0$ as $h \to 0$. Form the set

$$S_i^{(h)} = [\tau_i,\,\tau_{i+1}) \times \{x \in \Omega : v_i(x) \ge \mu\}\,, \tag{1.8}$$

where μ is some positive number $\beta_1 \le \mu \le \beta_2$.

Denote

$$S^{(h)} = \bigcup_{i=0}^{m-1} S_i^{(h)}\,,$$

where $d(S_*(\theta),\,S^{(h)})$ is the Lebesgue measure of the symmetric difference of sets S_*, $S^{(h)}$.

Theorem. If $h \to 0$ then the following is valid

$$|v^{(h)} - u_*|_{L^2(T;L^2(\Omega))} \to 0$$

$$d(S(\theta),\,S^{(h)}) \to 0 \, .$$

Remark 1.2. Similar to [12] one can obtain an estimate of reconstruction accuracy.

2. Consider an example. Let ϕ be a convex continuous function under the assumption of Section 1. Then the system (1.1) is equivalent to the equation

$$\frac{\partial y}{\partial t} = Ay + u + f(t,x),\ t\in T,\ x\in\Omega,\ y|_\Gamma = 0 \tag{2.1}$$

Here A is an elliptic coercive operator

$$Ay = \frac{\partial}{\partial x_j}\left(a_{ij}(x)\frac{\partial y}{\partial x_i}\right) - q(x)y,\ a_{ij} = a_{ji}, \tag{2.2}$$

$$a_{ij} \in L^\infty(\Omega),\ q \in L^\infty(\Omega).$$

For (2.1) consider a concrete variant of reconstruction problem [12].

Let Ω be a two-dimensional domain

$$0 < x_1 < \ell_1,\ 0 < x_2 < \ell_2;\ f = 0,\ q = 0$$

and

$$Ay = a^2 \cdot \partial^2 y/\partial x_1^2 + b^2 \cdot \partial^2 y/\partial x_2^2 .$$

For the sake of simplicity we confine the considerations to the case of reconstruction of location $G(t)$, $t\in T$. Let it be known a priori that the control being restored satisfying the inequality $|u(t,\cdot)|_{L^2(\Omega)} \le R$.

A closed ball in $L^2(\Omega)$ of radius R is taken as P. Then

$$v_i = [\varsigma(\tau_i) - z(\tau_i\,;\,t_0,y_0,v(\cdot))] \,/\, \alpha(h) \text{ if}$$

$$|\varsigma(\tau_i) - z(\tau_i\,;\,t_0,y_0,v(\cdot))|_{L^2(\Omega)} \le R\cdot\alpha(h),$$

$$v_i = R\cdot[\varsigma(\tau_i) - z(\tau_i\,;\,t_0,y_0,v(\cdot))] \,/\, |\varsigma(\tau_i) - z(\tau_i\,;\,t_0,y_0,v(\cdot))|_{L^2(\Omega)}, \text{ if}$$

$$|\varsigma(\tau_i) - z(\tau_i\,;\,t_0,y_0,v(\cdot))|_{L^2(\Omega)} > R\cdot\alpha(h).$$

For the considered variant of the problem the calculations were carried out for the following data

$$a^2 = b^2 = 0.1,\ \ell_1 = \ell_2 = 10,\ t_0 = 0,\ \theta = 1,\ R = 100,$$

$$y_0 = 0,\ \beta_1 = \beta_2 = 10,\ \delta(h) = h,\ \alpha(h) = \sqrt{h},\ h = 0.1 .$$

The motions of the dynamical system and the auxiliary model were calculated with the help of an explicit difference scheme with constant time step $\tau = \delta(h)$ and constant spatial steps γ_1 and γ_2 in x_1 and x_2 respectively.

The set $G(t_0)$ is depicted in Fig. 1 and Figs. 2 and 3 show the results of reconstruction of the set

$$G(t) = \{(x_1,x_2) : 0.01 \le x_1 \le 9.99, x_1(t,x_1) \le x_2 \le x_2(t,x_1)\} ,$$

where

$$x_1(t,x_1) = 3.5 + \cos(0.5\cdot x_1 - 5\cdot t) + 0.3\cdot\cos(5\cdot x_1 + t/h)\cdot\sin(3.2\cdot x_1 + t/h) ,$$

$$x_2(t,x_1) = 6.5 + \cos(0.5\cdot x_1 - 5\cdot t) + 0.3\cdot\cos(10\cdot x_1 + t/h)\times\sin(3.2\cdot x_1 + t/h) ,$$

at the moments $t = 0.5$, $t = 0.9$ respectively for

$$\gamma_1 = \gamma_2 = 10/16 .$$

The unknown set is reconstructed with the help of rectangles with centres in the mesh nodes and sides γ_1 and γ_2 parallel to axes x_1, x_2 respectively.

The author wishes to express gratitude to A.V. Kryazhimski, A.V. Kim, A.I. Korotki, V.I. Maksimov for valuable discussions and assistance, and also to A.M. Ustyuzhanin for help in computer simulation of the illustrative example.

References

[1] Osipov, Yu.S., Kryazhimski, A.V. Method of Lyapunov functions for problems of motion modelling. 4th Chetayev's Conference on Motion Stability, Analytical Mechanic and Control. Zvenygorod (USSR), 1982. Abstracts, p. 35 (in Russian).

[2] Kryazhimski, A.V., Osipov, Yu.S. On modelling of control in a dynamical system. Izv. Akad. Nauk USSR, Tech. Cybern. 1983. No. 2, pp. 51-60 (in Russian).

[3] Osipov, Yu.S., Kryazhimski, A.V. On dynamical solution of operator equations. Dokl. Akad. Nauk (USSR), 1983. Vol. 269, No. 3, pp. 552-556 (in Russian).

[4] Kurzhanski, A.B. Control and observation under uncertainty. Moscow, Nauka, 1977 (in Russian).

[5] Gusev, M.I., Kurzhanski, A.B. Inverse problems of dynamics of control systems. In: Mechanics and Scientific-Technical progress. Vol. 1, Moscow, Nauka, 1987 (in Russian).

[6] Kryazhimski, A.V., Osipov, Yu.S. Inverse problems of dynamics and control models. In: Mechanics and Scientific-Technical progress. Vol. 1, Moscow, Nauka, 1987, pp. 196-211 (in Russian).

[7] Kryazhimski, A.V. Optimization of the ensured result for dynamical systems. Proceedings of the Intern. Congress of Mathematicians, Berkeley (USA), 1986. pp. 1171-1179.

[8] Osipov, Yu.S. Control problems under insufficient information. Proc. of 13th IFIP Conference "System modelling and Optimization", Tokyo, Japan, 1987. Springer, 1988.

[9] Kryazhimski, A.V., Osipov, Yu.S. Stable solutions of inverse problems for dynamical control systems. Optimal Control and Differential Games, Tr. Matem. Inst. im. Steklova, USSR, 1988. Vol. 185, pp. 126-146 (in Russian).

[10] Maksimov, V.I. On dynamical modelling of unknown disturbances in parabolic variational inequalities. Prikl. Mat. Mekh., 1988. Vol. 52, No. 5, pp. 743-750 (in Russian).

[11] Kim, A.V., Korotki, A.I. Dynamical modelling of disturbances in parabolic systems. Izv. Akad. Nauk, USSR. Tekhn. Kibernet. (in Russian, to appear).

[12] Kim, A.V., Korotki, A.I., Osipov, Yu.S. Inverse problems of dynamics for parabolic systems. Prikl. Math. Mekh. (in Russian, to appear).

[13] Osipov, Yu.S. Dynamical reconstruction problem. 14th IFIP Conference, Leipzig, 1989.

[14] Krasovski, N.N., Subbotin, A.I. Game-theoretical control problems. Springer-Verlag, New York, 1987.

[15] Krasovski, N.N. Controlling of a dynamical system. Moscow, Nauka, 1985 (in Russian).

[16] Osipov, Yu.S. On theory of differential games for the systems with distributed parameters. Dokl. Akad. Nauk, SSSR, 1975. Vol. 223, No. 6 (in Russian).

[17] Osipov, Yu.S. Feed-back control for parabolic systems. Prikl. Mat. Mekh. 1977, Vol. 41, No. 2 (in Russian).

[18] Tikhonov, A.N., Arsenin, V.Ya. Solution of ill-posed problems. Wiley, New York, 1977.

[19] Duvaut, G., Lions, J.-L. Les inequations en mecanique et en physique. Dunod, Paris, 1972.

[20] Glowinski, R., Lions, J.-L., Tremolieres, R. Analyse numérique des inequations variationnelles. Dunod, Paris, 1976.

[21] Barbu, V. Optimal feed-back controls for a class of nonlinear distributed parameters systems. SIAM J. Contr. Opt., Vol. 21, No. 6, pp. 871-894.

[22] Brezis, H. Operateurs maximaux monotones et semigroupes de contractions dans les espaces de Hilbert. North-Holland, Elsevier, 1973.

[23] Kurzhanski, A.B., Osipov, Yu.S. On the problems of program pursuit. Izv. Akad. Nauk, USSR, Tech. Cybern., No. 3, 1970. (Translated as Engineering Cybernetics)

[24] Osipov, Yu.S. Inverse problems of dynamic. Report on 7th International Seminar, Tbilisi, 1988.

[25] Kurzhanski, A.B. Identification - a theory of guaranteed estimates. IIASA Working Paper WP-88-55, 1988.

[26] Kurzhanski, A.B., Khapalov, A.Yu. On the state estimation problem for distributed systems. Analysis and Optimization of Systems. Lecture Notes in Control and Information Sciences. Vol. 83, Springer-Verlag, 1986.

[27] Kurzhanski, A.B., Khapalov, A.Yu. Observers for distributed parameter systems. Control of Distributed Parameter Systems. Fifth IFAC Symposium. University of Perpignan, 1989.

[28] Kurzhanski, A.B., Sivergina, I.F. On noninvertible evolutionary systems: guaranteed estimation and the regularization problem, IIASA Working Paper, November 1989, forthcoming.

SENSITIVITY ANALYSIS OF SHALLOW SHELL WITH OBSTACLE

Murali Rao
Department of Mathematics
University of Florida
201 Walker Hall
Gainesville, FL 32611
USA

Jan Sokołowski
Systems Research Institute
Polish Academy of Sciences
ul. Newelska 6
01-447 Warszawa
POLAND

1. INTRODUCTION

In the present paper we provide some new results on the sensitivity analysis of variational inequalities. We shall consider the following obstacle problem for shallow shell [K-1].

Find an element $\omega=(w,u,v)\in K\subset H$ such that

$$\mathcal{A}(\omega,\phi - \omega) \geq \langle f,\phi - \omega\rangle, \quad \forall\phi\in K \tag{1.1}$$

where K is a convex , closed subset of the Sobolev space

$$H=H_0^2(\Omega)\times H_0^1(\Omega)\times H_0^1(\Omega)$$

$\mathcal{A}(.,.) : H\times H \to R$ is a bilinear form , $f \in H' = H^{-2}(\Omega)\times H^{-1}(\Omega)\times H^{-1}(\Omega)$ is a given element , and Ω is a given domain in R^2.

We shall prove that the solution ω of the variational inequality is directionally differentiable in the sence of Hadamard with respect to the right hand side f. To this end we use the concept of polyhedric convex set [H], [M], see Definition 1 below.

We refer the reader to [R-S-1] for related results in the scalar case , including the sensitivity analysis of the Kirchhoff plate with an obstacle, and state constrained optimal control problem [R-S-2] for elliptic equation. Some applications to the sensitivity analysis of optimization problems are provided in [S-1] - [S-4]. We use standard notation throughout the paper [A], [L-M].

2. DIFFERENTIABILITY OF PROJECTION ONTO K

First , we introduce the notation.

Convex set K is defined in the following way

$$K = \{ \phi = (\phi_1,\phi_2,\phi_3) \in H_0^2(\Omega)\times H_0^1(\Omega)\times H_0^1(\Omega) \mid \mathfrak{R}\phi \geq \psi \text{ in } \Omega \} \tag{2.1}$$

here $\psi = \psi(x)$, $x \in \Omega$, is the obstacle , and $\mathfrak{R}$ is the linear mapping

$$\mathfrak{R}\phi = \phi_1 - a_2\phi_2 - a_3\phi_3 \tag{2.2}$$

where $a_2 = \partial\psi/\partial x_1$, $a_3 = \partial\psi/\partial x_2$, here we assume for the sake of simplicity that the obstacle is sufficiently smooth , hence

$$\mathfrak{R}\phi \in H_0^1(\Omega) \text{ , for all } \phi \text{ in } H \tag{2.3}$$

Denote

$$H_0^1(\mathfrak{R};\Omega) = \{ \rho \in H_0^1(\Omega) \mid \rho = \mathfrak{R}\phi \text{ , for some } \phi \text{ in } H \} \tag{2.4}$$

the image of the mapping $\mathfrak{R}$ in the Sobolev space $H_0^1(\Omega)$.

Let us consider the metric projection P_K in H onto the set K.

We shall show that the set K is polyhedric [M] in the Sobolev space $H=H_0^2(\Omega)\times H_0^1(\Omega)\times H_0^1(\Omega)$ equipped with the scalar product

$$(\phi,\varphi) = (\phi,\varphi)_H = \int_\Omega (\Delta\phi_1\Delta\varphi_1 + \nabla\phi_2.\nabla\varphi_2 + \nabla\phi_3.\nabla\varphi_3)dx \tag{2.5}$$

and therefore the metric projection P_K is directionally differentiable in the sense of Hadamard [M]. In section 3 we use the result on directional differentiability of projection for the sensitivity analysis of variational inequality (1.1).

Let $T_K(\phi)$ denotes the tangent cone to K at $\phi \in K$. It is clear that $T_K(\phi)$ is the closure in the space H of the following convex cone

$$C_K(\phi) = \{\varphi \in H \mid \exists\ t > 0 \text{ such that } \phi + t\varphi \in K \} \tag{2.6}$$

For a given element $g \in H$, such that $\phi = P_K(g)$ let us define the following convex cone in the space H

$$S = T_K(\phi)\cap\mathfrak{H} \tag{2.7}$$

where

$$\mathfrak{H} = \{ \varphi \in H \mid (\phi,\varphi) = (g,\varphi) \} = [g - P_K(g)]^\perp \tag{2.8}$$

DEFINITION 1

The set K is polyhedric provided for any $g \in H$

$$S = \mathrm{cl}(C_K(\phi)\cap\mathfrak{H})$$

here cl stands for the closure. ■

First we derive the form of tangent cone $T_K(\omega)$ for any $\omega \in K$.

THEOREM 1.

Tangent cone $T_K(\omega)$ takes the form

$$T_K(\omega) = \{ \phi = (\phi_1,\phi_2,\phi_3) \in H_0^2(\Omega)\times H_0^1(\Omega)\times H_0^1(\Omega) \mid \mathfrak{R}\phi \geq 0 \text{ q.e. on } \Xi \} \tag{2.9}$$

where

$$\Xi = \{ x \in \Omega \mid \mathfrak{R}\omega(x) = \psi(x) \} \tag{2.10}$$

is the coincidence set.

Here q.e. means " quasi everywhere " with respect to the capacity of the space $H_0^1(\mathfrak{R};\Omega)$ equipped with the smallest norm for which the mapping $\mathfrak{R}$ is continuous . ■

PROOF OF THEOREM 1.

We assume that the coincidence set $\Xi \subset \Omega$ is compact. We denote by $\mathfrak{M}$ the following closed convex cone

$$\mathfrak{M} = \{ \phi \in H \mid \mathfrak{R}\phi \geq 0 \text{ , on } \Xi \} \tag{2.11}$$

It is clear that $C_K(\omega) \subset \mathfrak{M}$ hence $T_K(\omega) \subset \mathfrak{M}$. Let $V \in \mathfrak{M}$ be a given element and let ϕ_0 denote the orthogonal projection of V onto the convex cone $T_K(\omega)$, then

$$(\phi_0 - V,\phi) \geq 0 \text{ , for all } \phi \in T_K(\omega) \tag{2.12}$$

$$(\phi_0 - V,\phi_0) = 0 \tag{2.13}$$

We claim

$$(\phi_0 - V,\phi) = 0 \text{ , if } \mathfrak{R}\phi = 0 \tag{2.14}$$

Indeed if $\mathfrak{R}\phi = 0$ then $\pm\phi \in C_K(\omega)$ so that (2.14) follows from (2.12).

Define on $H_0^1(\mathfrak{R};\Omega)$ the positive linear map , well defined by (2.14),

$$Lv = (\phi_0 - V,\phi) \text{ , } v = \mathfrak{R}\phi \in H_0^1(\mathfrak{R};\Omega)$$

Then there is a positive Radon measure λ on Ω such that

$$Lv = \int_\Omega \mathfrak{R}\phi \, d\lambda = \int_\Omega v \, d\lambda$$

We claim that λ is concentrated on Ξ. Indeed if $\phi_0 \in C_0^\infty(\Omega\setminus\Xi)$ then clearly $\pm\phi = \pm(\phi_0,0,0)$ belongs to $C_K(\omega)$ so that from (2.12) it follows

$$\int_\Omega \phi_0 \, d\lambda = L\phi_0 = (\phi_0 - V,\phi) = 0$$

Finally in view of (2.13)

$$0 \leq (\phi_0 - V,\phi_0 - V) = - (\phi_0 - V,V) = - \int_\Omega \mathfrak{R}V \, d\lambda$$

Now λ is concentrated on Ξ, $\mathfrak{R}V \geq 0$ on Ξ so that the last quantity is non- positive. Hence we must have

$$\phi_0 = V$$

Since the element $V \in T_K(\omega)$ is arbitrary it follows that

$$T_K(\omega) = \mathfrak{M}$$

■

THEOREM 2.

Let $g \in H$ be a given element , denote $\phi = P_K g$, then the following condition is satisfied

$$T_K(\phi)\cap[\phi - P_K g]^\perp = \text{cl} \left(C_K(\phi)\cap[\phi - P_K g]^\perp \right)$$

condition is satisfied

$$T_K(\phi)\cap[\phi - P_K g]^{\perp} = \mathrm{cl}\ (\ C_K(\phi)\cap[\phi - P_K g]^{\perp})$$

therefore the convex cone K is polyhedric. ■

PROOF OF THEOREM 2.

We denote by $\Xi(\phi)$ the coincidence set

$$\Xi(\phi) = \{\ x \in \Omega\ |\ \Re\phi(x) = \psi(x)\ \}$$

There exists a nonnegative Radon measure ν such that $\nu(\Xi\setminus\Omega) = 0$, furthermore

$$[\phi - P_K g]^{\perp} = \{\ \phi \in H\ |\ \int_\Omega \Re\phi\ d\nu = 0\ \}$$

We have also

$$T_K(\phi) = \{\ \varphi = (\varphi_1,\varphi_2,\varphi_3) \in H_0^2(\Omega)\times H_0^1(\Omega)\times H_0^1(\Omega)\ |\ \Re\varphi \geq 0 \text{ q.e. on } \Xi(\phi)\ \}$$

We shall show that for any $V = (V_1,V_2,V_3) \in T_K(\phi)\cap[\phi - P_K g]^{\perp}$ the metric projection ϕ_0 of V onto the cone $\mathrm{cl}(C_K(\phi)\cap[\phi - P_K g]^{\perp})$ coincides with V i.e.,

$$V = \phi_0$$

Now

$$(\phi_0 - V,\varphi)_H \geq 0\ ,\ \text{for all } \varphi \in \mathrm{cl}(C_K(\phi)\cap[\phi - P_K g]^{\perp})$$

and there exists a nonnegative Radon measure λ such that

$$(\phi_0 - V,\varphi)_H = \int_\Omega \Re\varphi\ d\lambda$$

STEP 1

Let $\xi_1 \in C_0^\infty(\Omega\setminus F)$, $F = \mathrm{supp}\lambda$, $\psi_1 \geq 0$ on $\Xi(\phi)$, then

$$\xi = (\xi_1,0,0) \in C_K(\phi)\cap[\phi - P_K g]^{\perp}$$

hence

$$(\phi_0 - V,\xi)_H = \int_\Omega \Delta(\phi_{01} - V_1)\Delta\xi_1 dx \geq 0$$

We can apply the same argument as in the case of the space $H_0^2(\Omega)$ [R-S-1] in order to show that the latter inequality implies

$$\int_{\Omega\setminus F} (\ (\Delta(\phi_{01} - V_1))^2 + (\phi_{01} - V_1)^2)dx = 0$$

therefore

$$\phi_{01} = V_1\ ,\ \text{in } H_0^2(\Omega) \subset C(\bar{\Omega})$$

STEP 2

Measure λ is concentrated on $\Xi(\phi)$.

Let $\rho_1 \in C_0^\infty(\Omega\setminus\Xi(\phi))$ be a given element , denote $\rho = (\rho_1,0,0)$, then

$$(\phi_0 - V,\rho)_H = \int_\Omega \rho_1 d\lambda \geq 0 \text{ , for } \rho \in C_K(\phi)\cap[\phi - P_K g]^\perp$$

since $\pm\rho \in C_K(\phi)\cap[\phi - P_K g]^\perp$ it follows that

$$\int_\Omega \rho_1 d\lambda = 0 \text{ , for all } \rho_1 \in C_0^\infty(\Omega\setminus\Xi(\phi))$$

STEP 3

Since $cl(C_K(\phi)\cap[\phi - P_K g]^\perp)$ is a cone it follows that

$$(\phi_0 - V,\phi_0)_H = 0$$

therefore $\Re\phi_0 = \phi_{01} - a_2\phi_{02} - a_3\phi_{03} = 0$, λ a.e. thus

$$\phi_{01} = a_2\phi_{02} + a_3\phi_{03} \text{ , } \lambda \text{ a.e.}$$

We have

$$(\phi_0 - V,\phi_0 - V)_H = \int_\Omega \Re(\phi_0 - V)d\lambda =$$

$$\int_\Omega[(\phi_{01}-V_0) - a_2(\phi_{02}-V_2) - a_3(\phi_{03}-V_3)]d\lambda =$$

by STEP 1

$$-\int_\Omega(a_2\phi_{02} + a_3\phi_{03})d\lambda + \int_\Omega(a_2V_2 + a_3V_3)d\lambda =$$

$$-\int_\Omega\phi_{01}d\lambda + \int_\Omega(a_2V_2 + a_3V_3)d\lambda =$$

$$-\int_\Omega(V_0 - a_2V_2 - a_3V_3)d\lambda = -\int_\Omega \Re V d\lambda \leq 0$$

since $\Re V \geq 0$ on $\Xi(\phi)$, therefore $\phi_0 = V$. ■

3. SENSITIVITY ANALYSIS

Let $\Omega\subset R^2$ be a given domain with smooth boundary $\Gamma=\partial\Omega$. We shall consider the following variational inequality

find an element $\omega=(w,u,v)\in K\subset H$ such that

$$\mathcal{A}(\omega,\phi - \omega) \geq \langle f,\phi - \omega\rangle, \quad \forall\phi\in K \tag{3.1}$$

here we denote

$$\mathcal{A}(\omega,\omega)\equiv a\int_\Omega(\Delta w)^2dx + \int_\Omega[\varepsilon_{11}^2+\varepsilon_{22}^2+2\sigma\varepsilon_{11}\varepsilon_{22}+\frac{1}{2}(1-\sigma)\varepsilon_{12}^2]dx \tag{3.2}$$

where $\varepsilon_{11}=\partial u/\partial x_1 + k_1w$, $\varepsilon_{22}=\partial v/\partial x_2 + k_2w$, $\varepsilon_{12}=\partial u/\partial x_2 + \partial v/\partial x_1$,
$a > 0$, $1 > \sigma > 0$ are given constants ,

$$\langle f,\phi\rangle\equiv \int_\Omega(f_1\phi_1 + f_2\phi_2 + f_3\phi_3)dx \tag{3.3}$$

for all $\phi = (\phi_1,\phi_2,\phi_3) \in H_0^2(\Omega)\times H_0^1(\Omega)\times H_0^1(\Omega)$, $f = (f_1,f_2,f_3)$ is a given element in $H' = H^{-2}(\Omega)\times H^{-1}(\Omega)\times H^{-1}(\Omega)$.

We denote by $\omega = \Pi(f)$ the unique solution of variational inequality (3.1).

The coincidence set for the solution ω of variational inequality (3.1) takes the form

$$\Xi = \{ x \in \Omega \mid \Re\omega(x) = \psi(x) \}$$

Let us consider the mapping

$$\Pi : H' \ni f \to \omega \in K \subset H \tag{3.4}$$

We shall show that the mapping Π is directionally differentiable in the sence of Hadamard.

THEOREM 3.

For any $h \in H'$ and for $\varepsilon > 0$, ε small enough

$$\Pi(f + \varepsilon h) = \Pi f + \varepsilon\Pi' h + o(\varepsilon) \tag{3.5}$$

where $o(\varepsilon)/\varepsilon \to 0$, with $\varepsilon\downarrow 0$ in the norm of H , uniformly with respect to h on compact subsets of H'. The element $q = \Pi'(h)$ is given by the unique solution of the following variational inequality

find an element $q=(w,u,v)\in S\subset H$ such that

$$\mathcal{A}(\omega,\phi - \omega) \geq (h,\phi - \omega), \quad \forall\phi\in S \tag{3.6}$$

PROOF OF THEOREM 3.

We provide a general proof which shows that if the metric projection in the Hilbert space H is directionally differentiable then it is differentiable in any scalar product defined by a symmetric, coercive bilinear form, provided an auxiliary condition is satisfied. It seems that the same argument can be used in the case of non - polyhedric convex set.

Let $g(.): [0,\delta) \to H$ be given. Consider the family of variational inequalities

$$\omega(t)\in K: \quad \mathcal{A}(\omega(t),\phi - \omega(t)) \geq (g(t),\phi - \omega(t)), \quad \forall\phi\in K \tag{3.7}$$

then

$$\omega(t) = P_K(\omega(t)-A\omega(t)+g(t)) \tag{3.8}$$

where $A \in \mathcal{L}(H\to H)$ is linear bounded mapping defined by bilinear form $\mathcal{A}$

$$\mathcal{A}(\omega,\phi) = (A\omega,\phi)$$

Denote

$$z(t) = \omega(t)-A\omega(t)+g(t) \tag{3.9}$$

$$S = T_K(\omega(0))\cap[\omega(0) - z(0)]^{\perp}$$

$$= T_K(\omega(0))\cap\{ \phi \mid (A\omega(0),\phi) = (g(0),\phi) \}$$

hence

$$\omega(t) = P_K(z(t)) \tag{3.10}$$

and we have

$$\omega'(0) = P_S(z'(0)) \tag{3.11}$$

provided $z(.)$ is strongly differentiable at 0^+. In such a case for $t>0$, t small enough

$$\omega(t) = \omega(0) + t\omega'(0) + o(t) \tag{3.12}$$

We should show that for any given $z'(0)$ we can select $g'(0)$ in such a way that

$$z'(0) = \omega'(0) - A\omega'(0) + g'(0) \tag{3.13}$$

$$= P_S(z'(0)) - AP_S(z'(0)) + g'(0) \tag{3.14}$$

Let us denote by S^* the polar cone , then $P_S + P_{S^*} = I$ and for any element $\phi \in H$ we have

$$\phi = P_S\phi + P_{S^*}\phi , \tag{3.15}$$

$$(P_S\phi,P_{S^*}\phi) = 0 \tag{3.16}$$

It can be shown that the condition

$$(z'(0)- P_S(z'(0)) + AP_S(z'(0)),\phi) = (g'(0),\phi) = 0 , \tag{3.17}$$

for all $z'(0)\in H$

implies $\phi=0$. On the other hand the image $\mathrm{Im}\mathcal{Z}$ of the continuous mapping

$$\mathcal{Z} : H \ni z'(0) \rightarrow z'(0)- P_S(z'(0)) + AP_S(z'(0)) \in H \tag{3.18}$$

takes the form

$$\mathrm{Im}\mathcal{Z} = S^* + AS$$

It can be shown , see PROPOSITION 1 below , that

$$\mathrm{Im}\mathcal{Z} = H$$

We proved that for any element $g'(0)$ which belongs to the image of mapping (3.18) we have

$$\omega'(0) = P_S(z'(0)) = P_S(\omega'(0) - A\omega'(0) + g'(0)) \tag{3.19}$$

the derivative is uniform with respect to $z'(0)$ on compact subsets of the space H.

(3.19) is equivalent to the following variational inequality

$$\omega'(0)\in S: (\omega'(0) - (\omega'(0) - A\omega'(0) + g'(0)),\phi - \omega'(0)) \geq 0,$$

for all $\phi\in S$

hence

$$\omega'(0)\in S: \qquad \mathcal{A}(\omega'(0),\phi - \omega'(0)) \geq (g'(0),\phi - \omega'(0)\ , \ \forall\phi\in S \qquad (3.20)$$

and we have (3.20) for any right hand side $g'(0)$ in H , which completes the proof.

PROPOSITION 1

We have

$$\mathrm{Im}\mathcal{Z} = S^* + AS = H$$

PROOF

Let $V \in H$ and φ_0 denotes its projection onto $S^* + AS$

$$\varphi_0 \in S^* + AS : \ (\varphi_0 - V,\varphi - \varphi_0) \geq 0 \ , \ \forall\varphi \in S^* + AS$$

Now if $\varphi \in S^* + AS$ then

$$\varphi + \varphi_0 \in S^* + AS$$

so

$$(\varphi_0 - V,\varphi) \geq 0 \ , \ \forall\varphi \in S^* + AS \qquad (3.21)$$

Take $\varphi \in S^*$ then

$$(\varphi_0 - V,\varphi) \geq 0 \ , \ \forall\varphi \in S^*$$

i.e.

$$(0- (V - \varphi_0),\varphi -0) \geq 0 \ , \ \forall\varphi \in S^*$$

hence $P_{S^*}(V - \phi_0) = 0$, thus

$$V - \phi_0 \in S \qquad (3.22)$$

There exists the unique elements $v_0 \in S^*$, $w_0 \in S$ such that

$$\varphi_0 = v_0 + w_0$$

From (3.21) it follows that

$$(w_0,\varphi) + (v_0 - V,\varphi) \geq 0 \ , \ \forall\varphi \in S^*$$

but $w_0 \in S$ so $(w_0,\varphi) \leq 0$, $\forall\varphi \in S^*$ therefore

$$(v_0 - V,\phi) \geq 0 \ , \ \forall\varphi \in S^*$$

which means that $P_{S^*}(V - v_0) = 0$, thus

$$V - v_0 \in S$$

so $V = v_0 + w_0'$, for some $w_0' \in S$

Finally , from (3.21)

$$\forall\varphi \in S^* + AS : 0 \leq (\varphi_0 - V,\varphi) = (v_0 + w_0 - v_0 - w_0',\varphi)$$

$$= (w_0 - w_0',\varphi) = (0 -(w_0' - w_0),\varphi -0)$$

In particular the projection of $(w_0' - w_0)$ on $S^* + AS$ is 0 so

$$w_0' - w_0 \in S$$

hence

$$A(w_0' - w_0) \in AS$$

But

$$0 \le (w_0 - w_0', A(w_0' - w_0))$$

by coercitivity of bilinear form $\mathcal{A}(.,.)$

$$\le -\alpha(w_0 - w_0', w_0 - w_0') \le 0 \ , \quad \alpha > 0$$

hence $w_0 = w_0'$ i.e. $\varphi_0 = V$. ■

REFERENCES

[A-1] R.A. Adams *Sobolev Spaces* Academic Press, New York 1975

[B-O-S] M.P. Bendsoe, N. Olhoff and J. Sokołowski, Sensitivity analysis of problems of elasticity with unilateral constraints, *J. Struct. Mech.* 13(2) 1985 p.201-222

[H-1] A. Haraux, How to differentiate the projection on a convex set in Hilbert space. Some applications to variational inequalities. *J .Math. Soc. Japan* 29(4) 1977 p.615-631

[H-C] E.J. Haug and J. Cea, *Optimization of Distributed Parameter Structures* Vols. I and II. Sijthoff and Noordhoff, Alpen aàn den Rijn, The Netherlands 1981

[K-1] A.M. Khludnev, Existence and regularity of solutions of unilateral boundary value problems in linear theory of shallow shells. (Russian) *Differentsial'nye Uravneniya* 11 (1984) p.1968-1975

[K-2] A.M. Khludnev, Variational approach to contact problem of shallow shell with rigid body. (Russian) *Differentsial'nye Uravneniya s Chastnymi Proizvodnymi. Akad. Nauk SSSR, Sibirsk. Otdel., Novosibirsk,* 1981, no.2, p.109-114

[L] J. Lovisek, Optimal design of cylindrical shell with a rigid obstacle *Aplikace Matematiky* 34(1989), No.1,p.18-32.

[L-M] J.L. Lions and E. Magenes, *Problemes aux limites non homogenes.* Dunod, Paris, 1968

[M-1] F. Mignot, Controle dans les inequations variationelles elliptiques. *J. Funct. Anal.*(22)1976 p.25-39

[R-S-1] M. Rao and J. Sokołowski, Sensitivity of unilateral problems in $H_0^2(\Omega)$ and applications (to appear)

[R-S-2] M. Rao and J. Sokołowski, Shape sensitivity analysis of state constrained optimal control problems for distributed parameter systems. Lecture Notes in Control and Information Sciences Vol.114, Springer Verlag (1989) p.236-245

[S-1] J. Sokołowski, Sensitivity analysis for a class of variational inequalities,In: [H-C], p.1600-1609

[S-2] J. Sokołowski, Optimal control in coefficients of boundary value problems with unilateral constraints *Bulletin of the Polish Academy of Sciences, Technical Sciences*, vol.31, no.1-12, 1983 p.71-81

[S-3] J. Sokołowski, Sensitivity analysis of control constrained optimal control problems for distributed parameter systems. *SIAM J. Control and Optimization* (25)6, 1987 p.1542-1556

[S-4] J. Sokołowski, Shape sensitivity analysis of boundary optimal control problems for parabolic systems. *SIAM Journal on Control and Optimization*. 1988

[S-Z-1] J. Sokołowski and J.P. Zolesio, Shape sensitivity analysis of unilateral problems. *SIAM J. Math. Anal.* 5(18), 1987 p.1416-1437

A note on an interaction between penalization and discretization

Tomáš Roubíček

Institute of Information Theory and Automation
Czechoslovak Academy of Sciences
Pod vodárenskou věží 4, 182 08 Praha 8,
Czechoslovakia

The aim of this note is to investigate phenomena appearing when a state-constrained optimal control (or optimal shape design, etc.) problem governed by some differential equation is handled numerically. Then we are forced to approximate the problem on finite dimension spaces by some discretization method like finite diferences or finite elements, and simultaneously to cope with the state space constraints by some dual method – here we confine ourselves to the simplest one, namely to the penalty function method, but the augmented Lagrangean method will behave essentially by the same manner. By author's knowledge, an interaction between discretization and penalization has not been studied yet, except some investigations in soviet literature collected in the book by F.P.Vasilev [3] which does not deal directly with the dual treatment of the state constraints, however. Though the matter is not too complicated, it is perhaps worth mentioning briefly here because, by author's experience, all possible events are not sometimes realized well by those who use discretization with penalization simultaneuously.

As most of the phenomena appear already on an abstract level, we may begin with the following abstract optimization problem

$$(P) \qquad \begin{cases} \text{minimize } f(u) \text{ on } u \in U \\ \text{subject to } g(u) \in C \end{cases}$$

where $f : U \to \bar{R}$ is a cost function, $g : U \to Y$ a state operator, U a set of admissible controls, Y a space of states, and $C \subset Y$ a set of admissible states. From now on, we shall suppose controlability of (P), that is $g(U) \cap C \neq \emptyset$. After penalization (with a parameter $\varepsilon > 0$) and discretization (with a parameter $h > 0$) we get a family of unconstrained optimization problems, each of which can be written in an abstract form:

$$(P_\varepsilon^h) \qquad \text{minimize } f_\varepsilon^h(u) = f^h + \varepsilon^{-1} p(g^h(u)) \text{ on } u \in U^h,$$

where $f^h : U^h \to \bar{R}$, $g^h : U^h \to Y$ are an approximate cost function and state operator, respectively, $U^h \subset U$ is an internal approximation of the set of admissible controls, and $p : Y \to \bar{R}$ is an appropriate penalty function; for simplicity we suppose that p is so easy to be evaluated that it need not be approximated by some p^h, which is often case, indeed.

To simplify the problem as much as possible, we will assume the following, quite strong assumptions:

(1) U is compact, its topology being denoted by τ,

(2) Y is a metric space, ρ its metric, C its closed subset,

(3) f, g are continuous, $f > -\infty$,

(4) p is continuous, $p(C) = 0$, $p(Y \backslash C) > 0$,

(5) U^h is closed in U, f^h, g^h are continuous in the (relativized) topology τ,

(6) $U^{h_1} \subset U^{h_2}$ for $h_1 \geq h_2 > 0$, $\bigcup_{h>0} U^h$ is dense in U, and

(7) $f^h \to f$, $g^h \to g$ uniformly in the sense:

$$\forall \varepsilon > 0 \ \exists h_0 > 0 \ \forall 0 < h \leq h_0 \ \forall u \in U^h : \ |f^h(u) - f(u)| \leq \varepsilon, \ \rho(g^h(u), g(u)) \leq \varepsilon.$$

Note that the assumptions (1)–(5) obviously guarantee existence of a minimizer both of (P_ε^h) and of (P), which is, however, not too much important because all phenomena studied below appear also in more general setting of the problem where compactness (1) need not be used, cf. [2].

Though the assumptions (1)–(7) may seem quite powerful on a first look, they cannot ensure the convergence of the minima of (P_ε^h) to the minimum of (P) (and *a fortiori* the convergence of minimizers, either) if only $\varepsilon, h \searrow 0$, as shown by the following example.

Example 1. Consider a very simple situation: $U = [-1, 1]$, $Y = R$, $f(u) = g(u) = u$, $C = \{+1, -1\}$, $U^h = [-1 + h, 1]$, $f^h \equiv f$, $g^h \equiv g$ on U^h, and $p = 1 - |u|$. All the assumptions (1)–(7) are fulfilled trivially, and clearly $\min(P) = -1$, and $\operatorname{Argmin}(P) = \{-1\}$. On the other hand, it is easy to compute that, for $\varepsilon < h/2$, $\min(P_\varepsilon^h) = 1$ and $\operatorname{Armin}(P_\varepsilon^h) = \{1\}$, which shows that neither the minimum, nor the minimizer of (P_ε^h) converge respectively to the minimum or the only minimizer of (P) when $\varepsilon, h \searrow 0$ and $\varepsilon < h/2$, that means when ε tends to zero too quickly in comparison with h.

What the assumptions (1)–(7) can guarantee is only the existence of a stability criterion "$h \leq \eta(\varepsilon)$" under which the convergence is ensured:

Theorem 1. *Under the assumptions (1)–(7) there exists* $\eta : R^+ \to R^+$ *such that*

(8) $$\lim_{\varepsilon, h \searrow 0, h \leq \eta(\varepsilon)} \min(P_\varepsilon^h) = \min(P) \ , \quad \textit{and}$$

(9) $$\limsup_{\varepsilon, h \searrow 0, \ h \leq \eta(\varepsilon)} \operatorname{Argmin}(P_\varepsilon^h) \subset \operatorname{Argmin}(P) \ ,$$

where "limsup" has the usual meaning, i.e. it contains all τ*-cluster points of all chosen subnets.*

The proof is, in fact, contained as a part of the proof of Theorem 4.3 in [2] and will be thus omitted here (however, Theorem 4.3 there itself is stated in terms of so-called minimizing filters instead of the sets of minimizers, not supposing any compactness).

It should be emphasize that Theorem 1 has a little practical usage because it does not say anything about the stability criterion "$h \le \eta(\varepsilon)$" except its mere existence. The following Theorems 2 and 3 provide us with more information, the former one dealing even with the extreme situation when no stability criterion is needed:

Theorem 2. *If (1)-(7) are fulfilled and moreover*

(10) $\qquad C = \mathrm{cl}_Y \mathrm{int}_Y C$ *and* $g(U) \cap \mathrm{int}_Y C \ne \emptyset$ *and* ,

(11) $\qquad \forall$ *uniform neighbourhood* B *of* $g^{-1}(\mathrm{int}_Y C)$ $\exists \delta > 0 : \; g^{-1}(C_\delta) \subset B$

where "cl_Y*" and "*int_Y*" stand respectively for the closure and the interior in* Y *and* C_δ *for* δ*-neighbourhood of* C *in the metric* ρ *. Then (8) and (9) hold with* $\eta \equiv 1$*, that means the convergence is unconditional.*

Again, the proof is essentially contained in [2] as a part of the proof of Theorem 4.4 and will be omitted here.

Remark 1. The hypothesis (10) is particularly satisfied if Y is a linear metric space, C is convex with nonempty interior and $g(U) \cap \mathrm{int}_Y C \ne \emptyset$; then we come to the standard Slater constraint qualification. As for (11), it is particularly satisfied if g^{-1} is uniformly continuous, possibly in the Haussdorff sense provided g^{-1} is multivalued.

Unfortunately, (11) is typically not fulfilled in optimization problems for systems governed by differential equations where usually Y is a normed linear space with a norm strictly coarser than the corresponding energetic norm; e.g. $Y = L^2(.)$ while the energetic space is some Sobolev space $H^k(.)$ with $k > 0$. In such case we have to perform the analysis more in detail, introducing also the auxiliary penalized problem without any discretization:

(P_ε) $$\text{minimize } f_\varepsilon(u) = f + \varepsilon^{-1} p(g(u)) \text{ on } u \in U.$$

Theorem 3. *Let (1)-(7) be fulfilled and the following discretization error is known:*

(12) $$\forall h \le h_0 : |\min(P_\varepsilon^h) - \min(P_\varepsilon)| \le E(\varepsilon, h_0).$$

Then every $\eta : R^+ \to R^+$ *such that* $\lim_{\varepsilon \searrow 0} E(\varepsilon, \eta(\varepsilon)) = 0$ *will guarantee (8) and (9).*

The proof of (8) follows from the fact that $\min(P_\varepsilon)$ converges for $\varepsilon \searrow 0$ to $\min(P)$ and from the obvious estimate:

$$|\min(P_\varepsilon^h) - \min(P)| \leq E(\varepsilon, \eta(\varepsilon)) + |\min(P_\varepsilon) - \min(P)|$$

provided $h \leq \eta(\varepsilon)$. As soon as (8) is proved, (9) is ensured simply by standard compactness arguments.

Example 2. We outline a rather model situation dealing with an optimal distributed-control problem for a nonlinear elliptic equation to illustrate how Theorem 3 can be applied. Let Ω be a bounded, polyhedral domain in R^n, $\partial\Omega$ its boundary, $U = \{u \in L^\infty(\Omega); -1 \leq u(x) \leq 1$ for a.a. $x \in \Omega\}$, τ is the topology induced on U from $H^1(\Omega)^*$ (which obviously guarantees (1), "$*$" stands for the topological dual space), $Y = L^2(\Omega)$, and $g(u) = y \in H^1(\Omega)$ is the weak solution of the nonlinear boundary value problem:

$$\nabla(a(|\nabla y|)\nabla y) = u \text{ on } \Omega, \tag{13}$$

$$a(|\nabla y|)\frac{\partial y}{\partial \nu} + y = 0 \text{ on } \partial\Omega \tag{14}$$

with some nonlinearity $a(.)$ such that the function $\xi \mapsto a(\xi)\xi$ is uniformly increasing with a linear growth, ν is the outward unit normal to $\partial\Omega$. In other words, $g(u) = y$ should fulfil the integral identity:

$$\int_\Omega a(|\nabla y|)\nabla y \; \nabla v \; dx + \int_{\partial\Omega} yv \; dS = \int_\Omega uv \; dx \qquad \forall v \in H^1(\Omega).$$

Furthermore, let

$$f(u) = \int_{\partial\Omega} g(u) \; dS, \tag{15}$$

C be a closed subset of $L^2(\Omega)$, and $p(y) = \inf_{\tilde{y} \in C} \|y - \tilde{y}\|^2_{L^2(\Omega)}$. In view of the cost function (15) together with the boundary conditions (14), we can see that, speaking in terms of a heat-transfer interpretation, we are to choose heat sources distributed around Ω in order to minimize the heat flux through the boundary $\partial\Omega$ representing a lost of energy outside the domain Ω, subject to some constraints imposed on the heat sources and on the temperature distribution. Hence our model problem has a quite reasonable practical interpretation.

We discretize the problem (13)–(14) by a standard manner, using the finite element method (any numerical integration is not needed here). Let $\{\Upsilon_h\}_{h>0}$ be a regular family of triangulations of Ω, $U^h = \{u \in U; u$ is piecewise constant on $\Upsilon_h\}$, $V^h = \{y \in H^1(\Omega); y$ is piecewise linear on $\Upsilon_h\}$, $f^h \equiv f$ on U^h, and $g^h(u) \in V^h$ is the unique solution of the integral identity:

$$\int_\Omega a(|\nabla g^h(u)|)\nabla g^h(u) \; \nabla v \; dx + \int_{\partial\Omega} g^h(u)v \; dS = \int_\Omega uv \; dx \qquad \forall v \in V^h.$$

To derive the estimate of the type (12) we employ the following facts:

i) f, g, and p are Lipschitz continuous on their respective sets of admissible arguments.

ii) The rate-of-error estimates which are uniform with respect to the control are known:

$$\|g(u) - g^h(u)\|_{L^2(\Omega)} \le c\ h^\alpha \qquad \forall u \in U^h, \text{and}$$

$$|f(u) - f^h(u)| \le c\ h^\beta \qquad \forall u \in U^h.$$

If the regularity $g(u) \in H^2(\Omega)$ is valid, by [3] it is well known that $\alpha = 1$, and in the linear case (i.e. $a \equiv const. > 0$) even $\alpha = 2$. As for β, its expected value is $\frac{1}{2}$ (or $\frac{3}{2}$ in the linear case), but we shall see (cf. Remark 2) that its concrete value has no influence on mere convergence (8) and (9).

iii) The uniform approximation error estimate is known:

(16) $$\inf_{u^h \in U^h} \|u - u^h\|_{H^1(\Omega)^*} \le c\ h^\gamma \qquad \forall u \in U.$$

Let us outline the proof of (16). For $u \in L^\infty(\Omega)$ denote by $u^h \in U^h$ the function defined by $\int_\Delta u^h\ dx = \int_\Delta u\ dx$ for every simplex $\Delta \in \Upsilon_h$. It is easy to verify that $\|v - v^h\|_{L^2(\Omega)} \le const. h \|v\|_{H^1(\Omega)}$ for every $v \in H^1(\Omega)$. Realizing that $\langle u - u^h, v^h\rangle = 0$ because evidently $\int_\Delta (u - u^h)\ dx = 0$ and v^h is constant on Δ for every $\Delta \in \Upsilon_h$, we obtain the estimate $|\langle u - u^h, v\rangle| = |\langle u - u^h, v - v^h\rangle| \le const.(\|u\|_{L^2(\Omega)} + \|u^h\|_{L^2(\Omega)})\ h\ \|v\|_{H^1(\Omega)}$. Taking into account that $u, u^h \in U$ and the definition of the standard dual norm, we can see that $\|u - u^h\|_{H^1(\Omega)^*} \le 2\ const.\sqrt{\text{meas}\Omega}\ h$, and put $\gamma = 1$ in (16).

Now we will employ the facts i)–iii) to derive the estimate (12). Taking some $u \in \text{Argmin}(P_\varepsilon)$, by (16) we can find some $u^h \in U^h$ with $\|u - u^h\|_{H^1(\Omega)^*} \le (c+1)h^\gamma$. By i) we can then see that $f_\varepsilon(u^h) \le \min(P_\varepsilon) + (c+1)h^\gamma(L + \frac{L^2}{\varepsilon})$, where L stands for the common Lipschitz constant of f, g, and p. By ii) we come to

(17) $$\min(P_\varepsilon^h) \le f_\varepsilon^h(u^h) \le \min(P_\varepsilon) + (c+1)(L + \frac{L^2}{\varepsilon})h^\gamma + c\ h^\beta + \frac{cL}{\varepsilon}h^\alpha.$$

Conversely, let us take some $u \in \text{Argmin}(P_\varepsilon^h)$. By ii) we get immediately

(18) $$\min(P_\varepsilon) \le f_\varepsilon(u) \le \min(P_\varepsilon^h) + c\ h^\beta + \frac{cL}{\varepsilon}h^\alpha.$$

Joining (17) and (18), we come to the error estimate (12) with

$$E(\varepsilon, h) = Const.(h^\gamma + h^\beta + \frac{1}{\varepsilon}(h^\gamma + h^\alpha)).$$

Then by Theorem 3, for the stability criterion function η we can take arbitrary function

$$\eta(\varepsilon) = \varepsilon^q \qquad \text{with } q > \max(\frac{1}{\alpha}, \frac{1}{\gamma}).$$

Remark 2. Note that β has no influence to a freedom of the choice of η, which is due to the fact that we investigated only mere convergence of the problem (P_ε^h) to (P), not any rate of convergence. Note also that the optimal case is $\alpha = \gamma$, particularly the case $\alpha = 2$ has here the same efficiency as $\alpha = 1$.

Remark 3. It is known that without the compactness hypothesis (1), the penalized problem (P_ε) does not generally approximate the original problem (P), but some extended problem (roughly speaking, a "relaxed control" problem). In such case, our considerations are also well fitted to approach relaxed controls by solving numerically the problems (P_ε^h); cf. [2] for a general treatment of this idea.

References

[1] P.Ciarlet, *The Finite Element Method for Elliptic Problems*, North-Holland, Amsterdam, 1978.

[2] T.Roubíček, Constrained optimization: a general tolerance approach. *Apl. Mat.* (to appear).

[3] F.P.Vasiljev, *Methods of Solving Extremal Problems* (in Russian), Nauka, Moscow, 1981.

PROBLEMS OF MATHEMATICAL MODELLING OF NONLINEAR PROCESSES

A.A. Samarskii
Keldysh Institute of Applied Mathematics
USSR Academy of Sciences
Moscow

One of the most important fields of modern science involves such key notions as optimization, control and optimal control. They reflect an aim of research, connected with efficiency, because optimization of any process and its control implements (in a certain sense) an optimal process. Unfortunately, classical methods employed in the control theory are efficient only for a narrow range of problems with elementary mathematical models. These methods prove to be non-efficient for complex problems described, e.g., by nonlinear partial differential equations. Meanwhile, it is obvious that optimization and control should be performed by basing rather on a profound and confident knowledge of an object under investigation than on its simplified models of the "black box" type. Therefore, to study properties of the object that we want to control is a first priority task. This task is generally very difficult since all modern problems under study are growing in complexity and scale, some becoming global (ecology, climate, etc.). Their study requires a system approach. For complex systems of any nature we need a forecast of their evolution,a scientific base of decision-making and control. The forecast must contain not only qualitative but quantitative characteristics as well - values of parameters determining a state of the system.

An extensive introduction of computer simulation greatly has widened a scope of problems that can be studied with the use of computers. We can talk to-day about computational physics, computational mechanics and so on.

Let us note some specific features of mathematical modelling in physics:

- using complex systems of various nonlinear equations of mathematical physics as physical-mathematical models of real physical processes;

-co-existence of several processes with different space-time scales;

- a hierarchy of models differring in their involvement of physical effects;
- a large range of variation of physical parameters;
- close connection of physics and engineering;
- a necessity to obtain optimal quantitative characteristics of processes under investigation with ensured accuracy;
- control of physical experiments;
- a need to know coefficients ("constants") of a nonlinear medium;
- identification of models.

All said above may be referred to other subject fields, characterizing a complexity level of computing simulation. The latest experience teaches us that in a program of theoretical research preceding the development of new technologies and the design we should include the study of possible scenarios of emergency situations to work out the control methods preventing failures.

We need a new methodology and a new technology in science. The mathematical modelling is such an universal methodology, and the computer simulation is the new technology. The point is to replace an original object or process (under investigation, control and operation) by its mathematical model and to experiment with it on a computer by means of computational and logical algorithms. The computer simulation consists of the following stages: a mathematical model - a computational algorithm - a respective program complex - computations - an analysis of results. This is the whole cycle. If necessary it is repeated (with a new model or algorithm or program or input), i.e. the computer simulation has an iterative nature and it is carried out for solving a class of problems by basing on a hierarchy of models of different completeness and accuracy.

We may single out two stages of the computer simulation: (1) choosing and verification of mathematical models, and (2) forecast. The model quality and, hence, the forecast accuracy depend on an accuracy of assigning the medium properties, i.e. coefficients in the equations. The medium properties can be determined by means of the computer simulation, for example, using a quantum-mechanical model of atom as it occurs in the plasma physics problems.

The intellectual core of the computer simulation is the triad: "Model-Algorithm-Program". Therefore, the computer simulation optimization (the increase of adequacy, accuracy, efficiency, etc.) must be performed for the entire triad and not for its separate elements only. It opens great opportunities for improving the program packages when complex problems are being solved.

This suggets the following goals of the system and applied programming.

- The development of technologies for solving problems that should support the triad "Model-Algorithm-Program".

- The construction of integrated architectures of computer systems and software oriented to classes of problems, the conceptual unity of hardware and software, a high level of adaptivity (flexibility of structures) to specific applications.

- Introduction of automatic programming for generating intellectual packages in major knowledge fields.

- The construction of the programming base relying on the computer simulation methodology, the development of a unified base for computer facilities and software.

- The orientation of the programmers' training not only to studying programming languages but to learning of the mathematical modelling and computer simulation methodology and applications.

The mathematical modelling has great methodological potentialities and can easily be adapted to solving various problems. Its universal nature manifests itself in that: (1) the different processes can be described by the same models (for example, by differential equations of the same type), and (2) despite a great variety of problems in any field of science and technology there is a finite number of main or base problems for which the models can be constructed which belong to a given manifold. Therefore, it is necessary to concentrate on analysis of the problem classes by distinguishing the base problems for which respective triads can be constructed to act as the modules for complex problems of a given class. It is obvious that the mathematical modelling:

- combines the merits of traditional theoretical and experimental methods;

- allows safety tests of objects in extremal (e.g., emergency) conditions where the field experiments are either impossible or dangerous;

- provides integration of the research, development, design, management and operation stages within interdisciplinary methodology.

At the present time introducing the new methodology to the fields such as biology, economy, ecology, sociology and the humanities as well as many engineering fields is held back by a lack of adequate models, which requires respective research.

Incomplete information about an object cannot prevent using an ideology of the triad. In this case an information model and the simulation are used in combination with expert systems ("soft" simulation). Of great importance here is the problem of developing the models of decision-making on the basis of incomplete information with involvement of mathematical modelling methods and intellectual expert systems.

Main processes in nature and society are nonlinear. Having arisen in physics and engineering, the notion "nonlinearity" now claims the status of a phylosophical category. It is dialectic, has many folds and represents an intrinsic property of any complex process. Now a new "nonlinear" thinking and a new technology of knowledge based on ideas of nonlinear nature are required.

Let us list main properties of nonlinear systems. They are:

- absence of the superposition principle (knowledege of the behaviour of fragments does not determine the behaviour of the whole);
- absense of the scale similarity and, hence, restriction of traditional experimental approaches;
- nonuniqueness of limit states of the evolution systems and evolution paths to these states;
- impossibility of direct extrapolation of the nonlinear system evolution process in both space and time (phase transitions, jumps, bifurcations, etc.);
- strong sensitivity to perturbations, critical states, thresholds;
- generation of "catastrophic" regimes in the course of evolution, for example, regimes with peaking when the nonlinear system parameters infinitely grow in a finite time.

Nonlinearity gives rise to many difficulties, but it contains several variants of evolution including the one with accelerated processes. We should admit that the nonlinearity is not an exotica,but it is a norm, while the linearity is an idealization true only under some restrictions. Examples of nonlinear media are microworld and plasma; atmosphere and ocean; active biological and chemical media; semiconductors; social, economic and ecological structures. Their respective models contain the heat conduction equations with temperature-dependent coefficients, the gas dynamic equations, the MHD equations, the radiative gasdynamics equations, the chemical kinetics equations, the Cortevege de Vriz and Schrödinger type equations, etc.

The nonlinearity means not only difficulties and complication but it opens new opportunities for control. Let us give two examples where the nonlinearity gives rise to additional parameters (the peaking time and the fundamental length).

Example 1. We have the equation of the chemical kinetics type

$$\frac{dn}{dt} = \beta n^2 - \alpha n, \quad t>0, \alpha, \beta = Const > 0, \quad n(0) = n_0$$

Here three types of solution are possible:

(1) $n(t) = n_0 = \alpha/\beta = Const,$

(2) for $n_0 < \alpha/\beta$ the solution $n(t) \to 0$ as $t \to \infty$;

(3) for $n_0 > \alpha/\beta$ the solution $n(t)$ infinitely grows over a finite interval of time as $t \to t_f$ (the solution is called the regime with peaking).

Thus, unlike the linear case = 0 we have here a new parameter $t_f = t(\alpha, \beta, n_0)$ representing the time scale (the peaking time).

Example 2. A distributed system - the process of combustion in a medium with nonlinear heat conduction and the nonlinear volume energy release - is described by the equation

$$\frac{\partial T}{\partial t} = \frac{\partial}{\partial x}\left(T \frac{\partial T}{\partial x}\right) + T^2, \quad t > 0, \quad T(x,0) = T_0(x), \quad -\infty < x < \infty$$

One of the solutions to this system (at a special choice of) is given by the formula

$$T(x,t) = \begin{cases} (t_f - t)^{-1} \cdot \frac{4}{3} \cos^2 \frac{\pi x}{L_s}, & |x| < \frac{L_s}{2} \\ 0, & |x| > \frac{L_s}{2} \end{cases}$$

Here, along with the time parameter t_f (the peaking time) we have another parameter L_s characterizing the solution distribution in space. As is seen from the formula for T(x,t) the combustion is localized in the domain $|x| < L_s/2$ (on the fundamental length L_s), while the rest medium does not "sense" what is going on over the interval $|x| < L_s/2$.

Appearance of new parameters and new scenarios of behaviour allows new possibilities for control and optimization. It is very difficult to study complex nonlinear processes directly without preparation. It is absurd to carry out direct computer simulation for them without obtaining, even with elementary models, preliminary information about some qualitative and quantitative characteristics of the processes under study. It is well understood by physicists who widely use the method of simple models for studying complex objects. Simple nonlinear models prove to be very pithy.

Let us illustrate this statement about effectiveness of simple nonlinear models by an example on combustion the specific case of which was considered in example 2 given above.

Let the combustion process be described by the equation for the temperature T (x, t) [1] :

$$\frac{\partial T}{\partial t} = \frac{\partial}{\partial x}\left(K_0 T^{\sigma} \frac{\partial T}{\partial x}\right) + q_0 T^{\beta}, \sigma > 0, \beta > 1, t > 0,$$

$$T(x, 0) = T_0(x), \quad -\infty < x < \infty$$

where $K(T) = K_0 T^{\sigma}$ is the thermal diffusivity, $q(T) = q_0 T^{\beta}$ is the density of a heat source.

Investigations show that depending on the relationship between and three quite different temperature regimes are possible.

I. At $\beta = \sigma + 1$ we have the regime of heat localization (see Example 2 (for $\sigma = 1$, $\beta = 2$) on the fundamental length $L_S = 2\pi\sqrt{\frac{\sigma+2}{\sigma}\frac{q_0}{K_0}}$ Here the heat does not disperse, and the temperature rises in the domain $|x| < L_S/2$ in the regime with peaking (the S-regime).

II. If $\beta < \sigma + 1$, the heat infinitely disperses and at a certain time we come across a blow-up of temperature (i.e. it goes to infinity) over the entire space (the HS-regime). The heat localization is absent.

III. If $\beta > \sigma + 1$, the localized temperature structure of the LS-regime is formed, while an effective size of the temperature field is preserved and infinite temperature is achieved only at a single point.

As we see, this simple model has great potentialities, for example, it contains three types of combustion regimes. Further investigations show that in more complex models, for example, in models of media under compression we also come across three similar regimes with peaking.

While solving the classical problem on compression of an ideal gas by a piston, when pressure on the piston grows in the peaking regime, we also revealed three compression regimes. If the piston pressure is given by the law [3]

$$p(0, t) = p_0 (t_f - t)^n, \quad n < 0, \quad t < t_f$$

then:

(I) at $n = n_s = -2\gamma/(\gamma+1)$, $\gamma > 1$, we have the S-regime (a localized standing wave of nonshock compression);

(II) if $n > n_s$, there is compression without shock wave, the localization occurs at a point (the LS-regime);

(III) if $n < n_s$, the shock wave has a finite velocity, there is no localization (the HS-regime).

It is shown that at the compression with peaking the complex gasdynamic structures may arise with localized density and temperature extrema determined by the entropy distribution in a gas.

The localization and structure generation effects suggest new methods of processes control. For example, the localization effect may prove useful in some plasma physics and thermal chemistry problems [2] . There are many examples proving that simple nonlinear models are extremely pithy: generation of heat structures in a medium with nonlinear heat conduction; self-focusing of light described by the nonlinear Schrödinger equation, collapse of Langmuire waves, collapse in gasdynamics problems. Using more complex models and studying them by the computer simulation confirmed at least qualitative (sometimes quantitative) characteristics obtained in simple models. A preliminary knowledge (even crude) of the process nature allows more efficient use of numerical methods including the adaptive ones.

Now we shall give two examples of successful investigations of real complex nonlinear problems by using the computer simulation. In the both cases we solved the process optimization problem by using the input control. The first example: simulation of processes in targets for laser thermonuclear fusion (a spherical denterium-tritium target is radiated by a laser). It is required to heat and compress the target up to the parameters when the thermonuclear reaction would be possible. It is also required to find optimal parameters at which the energy release is maximal for a given laser energy. The process is rather complex: on the target periphery "a corona" is formed, while the interior ("the core") is compressed and heated. Without going into details of the mathematical model we note that it involves the gasdynamic motion, the two-temperature condition, the electron heat conduction, the radiation, the complex state equations, the kinetics of thermonuclear reactions, neutrons, alpha-particles, etc.

It has been found out that we should differentiate between two variants of the subtype of laser impulse applied to the target surface. In the first case the laser power negligibly changes over the impulse time. The second case [4] - when the laser radiation power changes in the peaking regime. Accordingly, two regimes are possible. In the first one a powerful shock wave arises, and hence, the following heat wave weakly changes the target density. This variant is not advantageous since a final target density is not large and the thermonuclear reaction energy is small too. In this case to ignite the target we must feed the energy of 10^9 . In the second regime the target is heated and compressed by a subsonic heat wave so that the thermonuclear reaction is possible at much lower laser energy (10^4 - 10^5 J). Thus, using the second regime allows reducing the laser energy by 4-5 orders of magnitude, which makes LTF a competitor among other projects. Here the regime control is made by changing the laser power in time.

The other example is optimization of technology of laser-plasma metal treatment [5] . An advantage of this technology is that metal in the nitrogen atmosphere is exposed to the reradiation of a plasma cloud formed near the metal surface rather than to the radiation of the laser itself. Plasma ions intensively infiltrate the surface layer of metal due to which it appreciably hardens. It is required to choose the treatment regime that would guard the metal against destruction by the laser radiation (as it occurs with metals directly exposed to the laser radiation).

It is an experimental fact that the laser beam destroys metals at the pressure of 30 atm and does not destroy at 100 atm. However, the pressure 100 atm is very high, and besides, the process in this case proved to be unstable. The problem was to understand the process physics by means of mathematical modelling and to determine a minimal pressure in nitrogen at which the metal is hardened without being destroyed. It was solved by computer simulation with different models involving the gasdynamics equations with chemical kinetics and two-dimensional equations of radiative gasdynamics. It turned out that the hardening could be achieved without metal destruction at the pressure 30 atm if a proper profile for the laser impulse power was chosen. In contrast to a standard time-constant impulse, its value should be decreased by an order of magnitude following a certain law. Then the plasma cloud protecting the metal from destruction by the laser impulse exists long enough and all the physical-chemical transformations necessary for the surface hardening had time to come near completion.

In this problem involving research and development as well as integration aspects the optimal regime of the laser-plasma unit is achieved by the laser impulse profile control.

Studying nonlinear processes requires a combination of all methods: analytical, numerical and experimental. Investigation of a complex nonlinear problem begins with a search for analytical (self-similar, asymptotical) solutions for simple models, special cases. Usually we have to develop new methods, to carry out mathematical analysis of resulting problems, to search for particular solutions. In many cases group-invariant methods based on transformations forming the Lee and Lee-Becklund groups prove to be efficient. The particular solutions obtained are used also for testing numerical techniques. Thus, using numerical methods within mathematical modelling stimulates the development of analytical methods. To solve nonlinear problems one needs the numerical methods that would correctly reflect basic properties of objects under investigation (for example, the conservation laws), be economical in computations and provide sufficient accuracy in a certain class of problems. Thus, the numerical methods optimal on a class of problems are required. An important property of computational algorithms (codes) is their adaptivity to a sought-for solution; in many cases it can be achieved by constructing adaptive grids. At the present time the codes with automatically computed dynamic adaptive grids are available.

The first-priority tasks in this field are:

- the development of methods for solving nonlinear grid equations arising at approximation of nonlinear differential equations, in particular, with systems of different types;
- the development of methods for solving multidimensional problems and problems with a strongly nonconjugate operator;
- the development of codes for solving grid equations with allowance for the computer architecture (vector processing, many- and multiprocessor computations, special processor computations, professional personal computers, etc.);
- the design of program complexes and packages for different classes of problems, different categories of users (student-engineer-scientist).

By using a new methodology great effort is to be taken to reconstruct the entire system of training of specialists in applied mathematics and informatics.

REFERENCES

1. Samarskii A.A., Galaktionov V.A., Kurdyumov S.P., Mikhailov A.P. Regimes With Peaking In Problems For Quasilinear Parabolic Equations, M. "Nauka", 1987, 476 pp.

2. Zmitrenko N.V., Kurdyumov S.P., Mikhailov A.P., Samarskii A.A. Localization of thermonuclear combustion in a plasma with electron heat conduction. Pis'ma v ZETF, 1977, v.26, vyp. 9, p.620-624.

3. Demidov M.A., Mikhailov A.P., Stepanova V.V. Localization and structures at gas compression in the regime with peaking. DAN SSSR. - 1985. - v.281, N 1.

4. Volosevich P.P., Degtyarev L.M., Kurdyumov S.P., Levanov E.I., Popov Yu.P., Samarskii A.A., Favorskii A.P. Process of superstrong compression of a substance and initiation of thermonuclear fusion by a powerful impulse of laser radiation. Fizika plazmy, 1976, N 2, N 6, p.883-897.

5. Samarskii A.A. Computational experiment in the problems of technology. Vestnik AN SSSR, 1984, N 3.

SOME RESULTS CONCERNING FREE BOUNDARY PROBLEMS SOLVED BY DOMAIN (OR SHAPE) OPTIMIZATION OF ENERGY

J.P. ZOLESIO
U.S.T.L.
Laboratoire de Physique Mathématique
Place E. Bataillon
34060 MONTPELLIER Cédex 2 - FRANCE

Key words : Incompressible Stoke's fluid, Free boundary, Eulerian semi derivative of Energy, necessary condition.

INTRODUCTION

For a linear incompressible stationary fluid we consider the energy $E(\Omega)$, Ω being the volume occupied by the fluid. The objective of the first section is to consider the minimization of $E(\Omega)$ and to underline the difficulties encountered to obtain existence results.

The objective of the second section is to study the necessary condition which are verified when is minimized or maximized with respect to the boundary Γ the minimum energy $E(\Omega)$ of a fluid.

Ω will denotes the volume of $\mathbb{R}^N$ occupied by the fluid, $N = 2$ or 3, Γ is a part of its boundary that we chose as a control parameter. The total boundary will be in general denoted by $\partial\Omega = \overline{\Gamma} \cup \overline{\Sigma}$, where Σ is an open part of $\partial\Omega$ an can be itself decomposed in several parts corresponding to several kind of boundary condition that will be imposed to the fluid on $\overline{\Sigma}$. Without loss of generality we shall suppose that Ω is contained in a fixed smooth domain D. We shall lose the generality when we suppose D bounded which is in some situations necessary to use some compact imbeddings of Banach spaces of functions defined over D. The speed of the fluid particle at a point x of Ω, at time t will be denoted by $u(t)\ (x) = u(t,x)$. When the flow is stationary, Γ does not depend on t, Γ is a stream line of the field u

1 - MINIMIZATION OF THE ENERGY

Assuming Γ smooth enough so that the normal field $n(x)$ (out going to Ω) is defined, say Γ of class C^1, we have for all t and $x \in \Gamma$

$$n(x).u(t,x) = 0 \qquad (1.1)$$

We consider the simples rehology as being the linear incompressible stoke's model

$$\operatorname{div} u(t,x) = 0 \quad \text{in} \quad \Omega \tag{1.2}$$

$$-\Delta u + \nabla p = f \quad \text{in} \quad \Omega \tag{1.3}$$

$$(\varepsilon(u).n)_\Gamma = g \quad \text{on} \quad \Gamma \tag{1.4}$$

where f is given in $L^2(D)$, g in $H^1(D)$, H is the mean curvature of the boundary Γ. Γ is assumed smooth enough in this strong formulation), $\varepsilon(u)$ is the displacement tenseur, $2\varepsilon(u) = Du + {}^*Du$, σ is a non negative given number

$$\sigma \geq 0$$

σ is the so called surface tension and $(\varepsilon(u).n)_\Gamma$ stands here for the tangential component of the vector $\varepsilon(u).n$ on Γ. The weak formulation of problem (1.1) - (1.4) is as follows : we introduce the Hilbert space

$$\mathcal{H}^1(\Omega) = \{u \in L^2(\Omega)^N, \quad \operatorname{div} \varphi = 0, \quad \varepsilon(u) \in L^2(\Omega)^{N^2}\} \tag{1.5}$$

Using the classical Green's Theorem :

$$\forall \varphi, \psi \in C^\infty(\overline{\Omega}, \mathbb{R}^N)$$

$$\int_\Omega \varepsilon(\varphi)..\varepsilon(\psi)dx + \int_\Omega \langle \operatorname{div}(\varepsilon(\varphi)), \psi\rangle d\Gamma = \int_{\partial\Omega} \langle \varepsilon(\varphi).n, \psi\rangle d\Gamma \tag{1.6}$$

it is obvious that $\varepsilon(\varphi).n$ is defined, as an element of $H^{-\frac{1}{2}}(\partial\Omega)^N = (H^{\frac{1}{2}}(\partial\Omega)^N)'$, as soon as $\varphi \in \mathcal{H}^1(\Omega)$ and $\operatorname{div}(\varepsilon(\varphi)) \in L^2(\Omega)^N$. (this fact derives from the extension of (1.6) when φ and ψ are in $H^1(\Omega)^N$, $\operatorname{div}(\varepsilon(\varphi)) \in L^2(\Omega)^N$). Let

$$\mathcal{H}^1_o(\Omega) = \{\varphi \in \mathcal{H}^1(\Omega) \mid \varphi.n = 0 \quad \text{on } \Gamma, \ \varphi = 0 \quad \text{on } \Sigma\}$$

The weak solution u of (1.1) - (1.4) is the unique minimizer in $\mathcal{H}^1_o$ of

$$E(\Omega) = \underset{u \in \mathcal{H}^1_o(\Omega)}{\mathrm{Min}} \quad \frac{1}{2}\int_\Omega \varepsilon(u)...\varepsilon(u)dx - \int_\Omega f.u\,dx + \int_\Gamma g.u\,d\sigma$$

From the well known Korn inequality on $\mathcal{H}^1(\Omega)$ $a(u,u) = (\int_\Omega \varepsilon(u)..\varepsilon(u)dx)^{\frac{1}{2}}$ is an equivalent norm to the $H^1_o(\Omega)$ norm.

The objective of this section is to minimize $E(\Omega)$ with respect to Ω. We shall first define $E(\Omega)$ when Ω is not a smooth domain in D but simply a measurable subset of D. The first objective is then to define $\mathcal{H}^1_o(\Omega)$ when Ω is a measurable subset of D.

In weak form the condition (1.1) on Γ can be written (assuming Σ empty)

$$U \in \mathcal{H}^1_o(D), \quad \forall \psi \in C^1(\overline{D})$$

$$\int_\Omega U.\nabla\psi\,dx = 0 \tag{1.7}$$

More precisely we have the

<u>Lemma 1.1</u>

Let Ω be a domain in D with Lipschitzian boundary Γ and U be an element in $\mathcal{H}^1_o(D)$ and $u = U_{/\Omega}$ (the restriction to Ω). Then u belongs to $\mathcal{H}^1_o(\Omega)$ if and only if U verifies (1.7)

Proof :

If u belongs to $H^1_o(\Omega)$ then $u.\nabla\phi = \operatorname{div}(\phi u)$ and by Stoke's theorem

$$\int_\Omega \operatorname{div}(\phi U)dx = \int_\Gamma U.n\ \phi\ dx = 0\ .$$

In view of the minimization of the Energy we introduce the perimeter of Ω relatively to D, $P_D(\Omega) = \|\nabla\chi_\Omega\|_{M^o(D)}$, where $M^o(D)$ is the Banach vector space of Bounded measure on D, see R. Temann [1], J.P. Zolesio [6], [7]

the problem $\operatorname{Inf}_{E\subset\Omega\subset D} E(\Omega) + \sigma P_D(\Omega)$ (1.8)

with g = 0 is then equivalent to the following are

$$\operatorname*{Inf}_{E\subset\Omega\subset D}\ \operatorname*{Inf}_{u\in H^1_o(D)}\ \sup_{\phi\in H^1_o(D)}\ e(\Omega,u,\phi) + \sigma P_D(\Omega) \qquad (1.9)$$

where

$$e(\Omega,u,\phi) = \int_\Omega (\tfrac{1}{2}|\varepsilon(u)|^2 - f_E.u + u.\nabla\phi)dx \qquad (1.10)$$

Let

$$Z(\Omega,u) = \sup_{\phi\in H^1_o(D)} e(\Omega,u,\phi) \qquad (1.11)$$

Lemma 1.2.

If $\chi_{\Omega_n} \to \chi_\Omega$ in $L^1(D)$ and $u_n \to u$ in $H^1_o(D)$ then $Z(\Omega,u) \leq \liminf_n Z(\Omega_n,u_n)$ (1.12)

Proof :

for each ϕ in $H^1_o(D)$ we have $e(\Omega_n,u_n,\phi) \to e(\Omega,u,\phi)$ as $n \to \infty$ then taking the supremum over ϕ we get (1.12).

The main difficulty arises in the fact that minimizing sequences (Ω_n,u_n) of problem (1.9) are such that χ_{Ω_n} is bounded in $BV(D,\mathbb{R}^N)$ but only $\chi_{\Omega_n}\ \varepsilon(u_n)$ is bounded in $L^2(D,\mathbb{R}^{N^2})$ (and $\varepsilon(u_n)$ is not apriori bounded in $L^2(D,\mathbb{R}^{N^2})$). One simple possibility to over come that difficulty is to consider for some given $\varepsilon,\varepsilon > 0$ the following perturbated problem

$$(P_\varepsilon)\quad \operatorname*{Inf}_{\substack{E\subset\Omega\subset D\\ u\in H^1_o(D)}}\ \sup_{\phi\in H^1_o(D)}\ e_\varepsilon(\Omega,u,\phi) + \sigma P_D(\Omega) \qquad (1.13)$$

with

$$e_\varepsilon(\Omega,u,\phi) = e(\Omega,u,\phi) + \varepsilon e(D,u,\phi) \qquad (1.14)$$

The boundary condition (1.4) will be changed for a transmission condition involving the parameter ε on Γ. The fluid is now in the whole domain D but Γ is an interface. For this situation we get the following existence result :

Proposition 1.3

There exist $(\Omega, u(\Omega)) \in \{\Omega,\ E \subset \Omega \subset D,\ P_D(\Omega) < \infty\} \times H_o^1(D)$ such that $u = U(\Omega)|_\Omega \in H_o^1(\Omega)$ and $\forall \psi \in H_o^1(D), \forall \Omega'$, $E \subset \Omega' \subset D, \forall u' \in H_o^1(D)$

$e_\varepsilon(\Omega, u, \psi) \leq \sup \{e_\varepsilon(\Omega', u', \varphi) \mid \varphi \in H_o^1(D)\}$

2 - NECESSARY CONDITION SOLVED BY STATIONARY DOMAINS

In this section we are concerned by the Eulerian derivative of the Energy functional with respect to the domain Ω. The Energy functional $E(\Omega)$ is associated to a stationary linear stoke's flow ; Ω is the volume of $\mathbb{R}^N$, $N = 2,3$, occupied by the flow and at each point x of Ω, $u(x) = (u_1(x), .., u_n(x))$ is the speed vector of the particle located at x. The fluid is assumed incompressible then we assume that

$$\operatorname{div}(u(x)) = 0, \quad x \in \Omega \tag{2.1}$$

On the boundary $\partial\Omega$ of Ω several boundary conditions are usualy imposed, the boundary being decomposed in several components $\partial\Omega = \overline{\Sigma} \cup \overline{\Gamma}$, u is given on Σ while Γ is the "free" part of the boundary. Without any loss of generality for the results we obtain in this section we set $\Sigma = \emptyset$ (empty set) and we introduce a forcing term $f \in L^2(D)^N$, E dans D being two given smooth bounded domains in $\mathbb{R}^N$, $E \subset D$ with $0 < \text{meas}(E) < \alpha < \text{meas}(D)$ we consider the set of admissible domains Ω in $\mathbb{R}^N$ such that

$$E \subset \Omega \subset D \tag{2.2}$$

$$\text{meas}(\Omega) = \alpha \tag{2.3}$$

The strong formulation of the Stoke's equation is then

$$-\Delta u + \nabla p = \chi_E f \quad \text{in } \Omega \tag{2.4}$$

where p is the preassure and χ_E the characteristic function of the set E.

$$u.n = 0 \quad \text{on } \Gamma \tag{2.5}$$

$$(\varepsilon(u).n)_\Gamma = 0 \quad \text{on } \Gamma \tag{2.6}$$

and the free boundary condition would be $\varepsilon(u).n.n$ prescribed or solution of some tangential problem on $\Gamma \setminus (\partial D \cup \partial E)$, see (2.39), (2.42).

For each smooth domain Ω we introduce the following Hilbert space

$$H(\Omega) = \{u \in H^1(\Omega)^N,\ \operatorname{div} u = 0,\quad u.n = 0 \text{ on } \partial\Omega\} \tag{2.7}$$

and the functionnal

$$J_\Omega(u) = \frac{1}{2} \int_\Omega \varepsilon(u)..\,\varepsilon(u)\,dx - \int_E f.u\,dx \tag{2.8}$$

From the Korn inequality we know that J_Ω is coercive on $H(\Omega)$ and that there exists a unique minimizing element u to J_Ω over $H(\Omega)$. It turns out from (1.6) that u is a weak solution to problem (2.1), (2.4), (2.5).

The energy functional is then

$$E(\Omega) = J_\Omega(u) = \text{Min}\,\{J_\Omega(v) \mid v \in H(\Omega)\} \tag{2.9}$$

Lemma 2.1

Let V be on admissible field, i.e. $V \in C^o([0, \varepsilon[, C^1(D,D))$ with $V.n = 0$ on ∂D. Let $\Omega_t = T_t(V)\ (\Omega)$ be the perturbated domain. Then the following transformation

$$u \to u_t = [(\det(DT_t))^{-1}\ DT_t.u] \circ T_t^{-1} \tag{2.10}$$

is a linear isomorphism from $H(\Omega)$ onto $H(\Omega_t)$

Proof :

It can be easily verified that

$$(\operatorname{div} u)\ \circ T_t = (\det(DT_t))^{-1} \operatorname{div}\ ((\det DT_t)\ (DT_t)^{-1} u \circ T_t) \tag{2.11}$$

and also that, n being the unitary normal field on Γ, out going to Ω, then on Γ_t we have

$$n_t = (\|M(DT_t).n\|^{-1}\ M(DT_t).n) \circ T_t^{-1} \text{ on } \Gamma_t \tag{2.12}$$

where $M(A)$ is the cofactor's matrix of A,

$M(A) = \det A\ {}^*A^{-1}$.

From (2.10), (2.12) we get, with $J_t = \det\ (DT_t)$, $(u_t.n_t) \circ T_t$ proportional to $<M(DT_t).n,\ DT_t.u>$, that is proportional to $<{}^*DT_t^{-1}.n,\ DT_t.u> = <n,\ DT_t^{-1}.DT_t u>$ then $(u_t.n_t) \circ T_t = a(t)\ u.n$, and it can be verified that for t small enough, $0 \leq t \leq \varepsilon$, the function a(t) is strictely positive on $\overline{D}$.

From (2.10) and (2.11) we get

$$(\operatorname{div} u_t) \circ T_t = J_t^{-1} \operatorname{div}\ [J_t (DT_t)^{-1}\ J_t^{-1}\ DT_t.u]$$

that is

$$(\operatorname{div} u_t) \circ T_t = J_t^{-1} \operatorname{div} u \tag{2.13}$$

Again, as J_t^{-1} is strictely positive (and continuous) on $\overline{D}$ we conclude the proof of the lemma

Corollary 2.2

u_t being defined by (2.10) we have

$$E(\Omega_t) = \operatorname*{Min}_{u \in H(\Omega)} J_{\Omega_t}(u_t) \tag{2.14}$$

We shall write $E(\Omega_t) = \operatorname*{Min}_{u \in H(\Omega)} F(t,u)$ (2.15)

with $F(t,u) = J_{\Omega_t}(u_t)$

$$= \frac{1}{2} \int_{\Omega_t} |\varepsilon[(J_t^{-1}\ DT_t.u) \circ T_t^{-1}]\ |^2\, dx - \int_E <f, (J_t^{-1}\ DT_t.u) \circ T_t^{-1}> dx \tag{2.16}$$

We assume now that Ω and f are smooth enough so that u_t, $t \geq 0$, is smooth, say $u_t \in H^2(\Omega_t)^N \cap H(\Omega_t)$. In that case the differentiation with respect to the parameter t of a minimum results applies(see J.P. Zolésio [4], M.C. Delfour, J.P. Zolésio [5], and we get from (2.15)

$$dE(\Omega; V) = \frac{d}{dt} E(\Omega_t)\Big|_{t=o} = \frac{\partial}{\partial t} F(0,u) \tag{2.17}$$

Then we concentrate our study on the calculation of the term $\frac{\partial}{\partial t} F(0,u)$, u being assumed smooth. From (2.16) we know that this derivative will involves two integrals a boundary integral and a volume integral over Ω. Only the first one requires smoothness of u. In bat form the necessary condition associated to any stationary domain will immediately turn to be a free boundary condition in strong version.

$$\frac{\partial}{\partial t} F(0,u) = \int_\Omega \varepsilon(u)..\varepsilon(\partial u)dx - \int_\Omega f\ \partial u\ dx + \int_\Gamma \frac{1}{2}\ |\varepsilon(u)|^2\ V(0).n\,d\Gamma - \int_\Gamma f.u\ V(0).n\,d\Gamma, \tag{2.18}$$

where $\partial u = \frac{\partial}{\partial t} u_t \Big|_{t=o} \in H^1(\Omega)^N$ (2.19)

is the element of $H^1(\Omega)^N$ such that $u_t = u + t\ \ \partial u + O(t)$

where $O(t) \in H^1(\Omega)^N$, $\|O(t)\| / t \to o$, $t \to o$.

We have the following characterization

Lemma 2.3

$$\partial u = -\ \mathrm{div}\ V\ u + DV.u - Du.V \tag{2.19}$$

Proof :

(2.19) derives from the classical results, see J.P. Zolésio [2], [4], J. Sokolowski - J.P. Zolésio [3], under smoothness assumption we have

$$\frac{d}{dt} f(t,T_t(x))|_{t=o} = \frac{\partial}{\partial t} f(0,x) + \nabla_x f(0,x).V(0,x)$$

and $\left(\frac{d}{dt} J_t\right)_{t=o} = \mathrm{div}\ V(0)$; $\left(\frac{d}{dt} DT_t\right)_{t=o} = DV(0)$

Remark 2.4

As the derivatives with respect to t and to the space variable x commutes we get

$$\left(\frac{\partial}{\partial t} \varepsilon(u_t)\right)_{t=o} = \varepsilon(\partial u) \tag{2.20}$$

Corollary 2.5

Let N_o be a smooth unitary extension of the normal field n ; If $\Gamma = \partial\Omega$ is of class $C^k, k \geq 1$, there exists such a $N_o \in C^{k-1}(\mathcal{U})$, $\mathcal{U}$ being a neighborhood of Γ. Consider speed vector fields V such that in $\mathcal{U}$ we have

$$V(t,x) = v(t,x)\ N_o(x) \tag{2.21}$$

Then the normal component of ∂u on Γ is given by

$$\partial u.n = \mathrm{div}_\Gamma\ (vu) \tag{2.22}$$

where $\mathrm{div}_\Gamma(\ \ )$ is the tangential divergence, see J.P. Zolésio [2], J. Sokolowski - J.P. Zolésio [3], defined by, e being a vector field on Γ, E any extension .f e,

$$\mathrm{div}_\Gamma\ e = \mathrm{div}\ E - \langle\varepsilon(E).n,n\rangle \tag{2.23}$$

(It turns out that the right hand side of (2.23) is independant on the choice of extension E defined itself on an arbitrary neighborhood $\mathcal{U}$).

<u>Proof</u> :

Let V be any admissible vector field ; from (2.19), (2.5) we get

$\partial u.n = \langle DV(0).u,n\rangle - \langle Du.V,n\rangle$

But when V takes the form (2.21) we get, see J.P. Zolésio [2],

$DV(0) = v(0)\, DN_o + n.{}^*\nabla v(0)$ on Γ

and then

$\partial u.n = \nabla_\Gamma v(0).u + v(0)\, \mathrm{div}_\Gamma\, u = \mathrm{div}_\Gamma\, (v(0)u)$

<u>Remark 2.6</u>

As u is a smooth divergence free field, from (2.23) and (2.1) we get (writting (2.1) on the boundary)

$$\mathrm{div}_\Gamma\, (u) = - \langle\varepsilon(u).n,n\rangle \tag{2.24}$$

We adopt the fluid mechanic notation $\varepsilon(u).n.n$ for this term.

Finely we obtain a first expressicn for the Eulerian semi-derivative of the Energy $E(\Omega)$:

<u>Proposition 2.7</u>

The speed field V is given verifying (2.21), then

$$dE(\Omega;V) = \int_\Omega (\langle -\Delta u - f,\ \partial u\rangle_{\mathbb{R}^N} dx + \int_\Gamma (\langle\varepsilon(u).n,\ \partial u\rangle_{\mathbb{R}^N} + \frac{1}{2}\, \varepsilon(u)..\varepsilon(u) v(0) - f.uv(0))d\Gamma \tag{2.25}$$

Using now the problem (2.1), (2.4), (2.5), (2.6) whose u is assumed to be the strong solution we get from (2.25)

<u>Corollary 2.6</u>

$$\begin{aligned} dE(\Omega;V) = & -\int_\Omega \partial u\, \nabla p\, dx - \int_\Gamma \mathrm{div}_\Gamma (u)\, \mathrm{div}_\Gamma\, (v(0)u)\, d\Gamma \\ & + \int_\Gamma \frac{1}{2}\, \varepsilon(u)..\, \varepsilon(u)\, v(0)\, d\Gamma - \int_\Gamma f.u\, v(0)\, d\Gamma \end{aligned} \tag{2.26}$$

<u>Proof</u> :

from (2.6) we get $\varepsilon(u).n = \varepsilon(u).n.n\ n$ on Γ, then from (2.23), (2.24) ;

$$\varepsilon(u).n = - \mathrm{div}_\Gamma (u)\, n \quad \text{on} \quad \Gamma \tag{2.27}$$

using (2.4), (2.27) and (2.22) in (2.25) we get (2.26)

<u>Lemma 2.7</u>

Assuming the preassure smooth enough, we have

$$\int_\Omega \partial u. \nabla p\, dx = \int_\Gamma p\, \mathrm{div}_\Gamma\, (v(0)u)\, d\Gamma \tag{2.28}$$

<u>Proof</u> :

From Lemma 2.1 we know that the element u_t belongs to $H(\Omega_t)$, then $\mathrm{div}\,(u_t) = 0$ in Ω_t . Taking the derivative with respect to t, which commutes with the divergence operator we get, at t=o, $\mathrm{div}\,(\partial u) = 0$ in Ω. Then $\partial u\, \nabla p = \mathrm{div}\,(p\, \partial u)$ and from Stoke's formula we get.

$$\int_\Omega \partial u.\nabla p \, dx = \int_\Gamma p \, \partial u.n \, d\Gamma \qquad (2.29)$$

The expression (2.28) derives from (2.29) and (2.22)

Corollary 2.8

Assuming the Speed field V such that (2.21) is verified and assuming Γ, u and p smooth enough we get the Eulerian semi-derivative as expected by the Structure Theorem (see J.P. Zolésio [2]), i.e. as a boundary expression :

$$dE(\Omega;V) = -\int_\Gamma (p+\mathrm{div}_\Gamma(u))\,\mathrm{div}_\Gamma(v(0)u)\,d\Gamma + \frac{1}{2}\int_\Gamma (\varepsilon(u)..\varepsilon(u) - f.u)\,v(0)\,d\Gamma \qquad (2.30)$$

Lemma 2.9

Assuming Γ,p and u smooth enough

$$\int_\Gamma u\,\mathrm{div}_\Gamma(u)\ \nabla_\Gamma v(0)\,d\Gamma = -\int_\Gamma \mathrm{div}_\Gamma[(\mathrm{div}_\Gamma(u)).u]v(0)\,d\Gamma \qquad (2.31)$$

Proof :

(2.31) directely derives from the by parts integration formula on Γ, see J.P. Zolésio [2], J. Sokolowski, J.P. Zolésio [3]. It must be noticed that as u.n = 0 the mean curvature H does not occur in (2.31) ; also the speed v must be zero at the boundary of Γ, for short we assume here that the boundary Γ has no boundary (it is a compact manifold)

The preassure on the free surface Γ of the fluid is in many example a constant. for example the atmospheric preassure. As p is defined up to a constant we shall now assume that p is zero on Γ, then

Proposition 2.10

Assuming Γ, u smooth enough and p=0 on the boundary we have

$$dE(\Omega;V) = \int_\Gamma (\nabla_\Gamma(\mathrm{div}_\Gamma u).u + \frac{1}{2}\,\varepsilon(u)..\varepsilon(u) - f\,.\,u)v(0)d\Gamma \qquad (2.32)$$

Proof :

we have $\mathrm{div}_\Gamma(a\,u) = a\,\mathrm{div}_\Gamma\,u + (\nabla_\Gamma a).u$ (2.33)

then from (2.30) and (2.31) we get (with p=0 on Γ) :

$$dE(\Omega;V) = +\int_\Gamma \mathrm{div}_\Gamma\,(u\,\mathrm{div}_\Gamma u)\,v(0)\,d\Gamma - \int_\Gamma (\mathrm{div}_\Gamma u)^2\,v(0)\,d\Gamma + \int_\Gamma (\frac{1}{2}\,\varepsilon(u)..\ \varepsilon(u) - f.u)\,v(0)\,d\Gamma$$

and with (2.33)

$$dE(\Omega;V) = \int_\Gamma ((\mathrm{div}_\Gamma\,u)^2 + u.\nabla_\Gamma(\mathrm{div}_\Gamma u) - (\mathrm{div}_\Gamma u)^2)\,v(0)\,d\Gamma + \ldots$$

and we get (2.32)

Remark 2.11

The expression $u.\nabla_\Gamma(\mathrm{div}\,u) = u.\nabla(\mathrm{div}\,u)$ that we got in (2.32) is in fact a material derivative, not with respect to the speed field V that we introduced to generate virtual deformations of the domain Ω, but with respect to the physical speed vector u(x) of the fluid. Let $X_o \in \Gamma$ be given and set

$$\frac{d}{dt}\, x(t,X_o) = u(x(t,X_o)), \quad x(0,X_o) = X_o \tag{2.34}$$

we introduce the flow Transformation $T_t(u)$ and

$$\gamma(t) = (\mathrm{div}_\Gamma u)\ (x(t,X_o)) = (\mathrm{div}_\Gamma u) \circ T_t(u)\ (X_o) \tag{2.35}$$

where u is the solution of the fluid problem (2.1), (2.4), (2.5), (2.6) with $p=0$ on the Boundary, Γ and u assumed to be smooth enough, we get (with $x = x(t,X_o)$ for short) :

$$\frac{d}{dt}\, \gamma(t) = \nabla_\Gamma (\mathrm{div}_\Gamma u)\ (x(t)).u(x(t)) \tag{2.36}$$

that is

$$\frac{d}{dt}\, (\mathrm{div}_\Gamma u) \circ T_t(u) = (\nabla_\Gamma\ \mathrm{div}\ u) \circ T_t(u).u \circ T_t(u) \tag{2.37}$$

Proposition 2.12. (Necessary optimality condition)

Let Ω be a smooth stationary domain of the Energy functional :

$$\forall\, V,\ \text{admissible field},\ dE(\Omega;V) = 0 \tag{2.38}$$

Then $\varepsilon(u).n.n = -\mathrm{div}_\Gamma u$ on Γ solves (the tangential differential equation), $\forall x_o \in \Gamma$,

$$[\varepsilon(u).n.n](x(t)) = (\varepsilon(u).n.n)\ (x_o) + ct - \int_o^t (\tfrac{1}{2}\, |\varepsilon(u)|^2(x(s)) + f(x(s)).u(x(s)))ds \tag{2.39}$$

where $|\varepsilon(u)|^2 = \varepsilon(u)..\ \varepsilon(u) = \sum_{i,j} (\varepsilon_{ij})^2$, C is a constant.

Proof :

We assume that Ω is a stationary domain for the functional E in the set of all admissible domain with prescribed measure. Then the field V(t,.) has to be chosen with free divergence so that the measure $\mathrm{meas}(\Omega_t) = \mathrm{meas}(\Omega)$ is given, from Stoke's formula, that is to say that the normal component v(0) of the field V(0)

$$\text{on } \Gamma \text{ verifies } \int_\Gamma v(0,x)\ d\Gamma(x) = 0 \tag{2.40}$$

From (2.40), (2.32) and (2.37) we get

$$\nabla_\Gamma(\mathrm{div}_\Gamma u).u + \frac{1}{2}\, |\varepsilon(u)|^2 - f.u = C \quad \text{on} \quad \Gamma$$

$$\frac{d}{dt}\, [(\mathrm{div}_\Gamma u) \circ T_t(u)] = C - \frac{1}{2}\, |\varepsilon(u)|^2 \circ T_t(u) - (f.u) \circ T_t(u) \tag{2.41}$$

where C is a constant deriving from (2.40), as we have the orthogonality to a closed subspace.

Remark 2.13

If the force f is zero on Γ, from (2.39) we get that $t \to (\varepsilon(u).n.n)\ (x(t))$ is monotoniquely decaying when the volume of the fluide is not prescribed.

Remark 2.14

If the volume of the domain Ω is not prescribed then the constant $C = 0$.

We consider now the situations involving the surface tension σ. We introduce

$$E_\sigma(\Omega) = E(\Omega) + \sigma P_D(\Omega) \tag{2.42}$$

where $P_D(\Omega)$ is the perimeter of Ω relatively to D.

Proposition 2.15

Let Ω be a smooth stationary domain for the functional E_σ, then, assuming $u = u(\Omega)$ smooth enough and the preassure $p = 0$ on Γ, the term $\varepsilon(u).n.n$ solves the following problem : $n_o \in \Gamma$, $x(t) = x_o + \int_o^t u(x(s))ds$,

$$[\varepsilon(u).u.n](x(t)) = (\varepsilon(u).n.n)(x_o) + ct$$
$$- \int_o^t [\sigma H(x(s)) + \frac{1}{2} |\varepsilon(u)|^2(x(s)) + f(x(s)).u(x(s))]ds \tag{2.43}$$

Where $H(x)$ is the mean curvature of the surface Γ at point x.

Proof :

When Ω is a smooth domain of class C^2 we have $P_D(\Omega) = \int_\Gamma d\Gamma$, where $\Gamma = \partial\Omega \setminus \partial D$ and then the Eulerian derivative of $P_D(\Omega)$, with an admissible field V such that $V = 0$ on $\partial D \cap \partial\Omega$, is given by, see J.P. Zolésio [7], [6]

$$dP_D(\Omega; V) = \int_\Gamma H\, V(0).n\, d\Gamma \tag{2.44}$$

where H is the mean curvature of Γ and n is the out going to Ω unitary normal field on Γ. Then (2.43) derives from (2.39) and (2.44).

CONCLUSION

This short study intends to underline that minimizing (or maximizing) the energy term $E(\Omega)$ with respect to the boundary Γ does not leads to the physical free boundary (as it is true for perfect fluid, for example in hydrodynamic, see Zolésio [ ]) but to the tangential problems (2.39), (2.41) or (2.43)

REFERENCES

[1] R. Teman Problèmes de Mathématiques en Plasticité, Dunod, 1984, Paris

[2] J.P. Zolésio The material (or Speed) method for Shape optimization in "optimization of Distributed Parameters", E. HAUG, J. CEA eds., Sijthof and Noordhoff 1980, p.p. 1152-1194.

[3] J. Sokolowski - J.P. Zolésio Introduction to shape optimization, Springer Verlag, 1990, in the Série Computational Mathematics.

[4] J.P. Zolésio - Shape sensitivity Analysis of reapeted Eigenvalues, in the same book then [2]

[5] M.C. Delfour - J.P. Zolésio Shape Sensitivity Analysis Via Min Max Differentiability, SIAM J. on Control and Optimization, 26, pp. 834-862 (1988)

[6] J.P. Zolésio, Existence and Uniqueness Results for Domain Variational Bernoulli Like Free Boundary Problem, to appear

[7] J.P. Zolésio, Existence Results for free boundary Problems, in "Distributed Parameter Systems", F. KAPPEL, K. KUNISH, W. SHAPACKER, eds., L.N.C.I.S 102, pp 333-343, 1986

Lecture Notes in Control and Information Sciences

Edited by M. Thoma and A. Wyner

Vol. 97: I. Lasiecka/R. Triggiani (Eds.)
Control Problems for Systems
Described by Partial Differential Equations
and Applications
Proceedings of the IFIP-WG 7.2
Working Conference
Gainesville, Florida, February 3-6, 1986
VIII, 400 pages, 1987.

Vol. 98: A. Aloneftis
Stochastic Adaptive Control
Results and Simulation
XII, 120 pages, 1987.

Vol. 99: S. P. Bhattacharyya
Robust Stabilization Against
Structured Perturbations
IX, 172 pages, 1987.

Vol. 100: J. P. Zolésio (Editor)
Boundary Control and Boundary Variations
Proceedings of the IFIP WG 7.2 Conference
Nice, France, June 10-13, 1987
IV, 398 pages, 1988.

Vol. 101: P. E. Crouch,
A. J. van der Schaft
Variational and Hamiltonian
Control Systems
IV, 121 pages, 1987.

Vol. 102: F. Kappel, K. Kunisch,
W. Schappacher (Eds.)
Distributed Parameter Systems
Proceedings of the 3rd International Conference
Vorau, Styria, July 6–12, 1986
VII, 343 pages, 1987.

Vol. 103: P. Varaiya, A. B. Kurzhanski (Eds.)
Discrete Event Systems:
Models and Applications
IIASA Conference
Sopron, Hungary, August 3-7, 1987
IX, 282 pages, 1988.

Vol. 104: J. S. Freudenberg/D. P. Looze
Frequency Domain Properties of Scalar
and Multivariable Feedback Systems
VIII, 281 pages, 1988.

Vol. 105: Ch. I. Byrnes/A. Kurzhanski (Eds.)
Modelling and Adaptive Control
Proceedings of the IIASA Conference
Sopron, Hungary, July 1986
V, 379 pages, 1988.

Vol. 106: R. R. Mohler (Editor)
Nonlinear Time Series and
Signal Processing
V, 143 pages, 1988.

Vol. 107: Y. T. Tsay, L.-S. Shieh, St. Barnett
Structural Analysis and Design
of Multivariable Systems
An Algebraic Approach
VIII, 208 pages, 1988.

Vol. 108: K. J. Reinschke
Multivariable Control
A Graph-theoretic Approach
274 pages, 1988.

Vol. 109: M. Vukobratović/R. Stojić
Modern Aircraft Flight Control
VI, 288 pages, 1988.

Vol. 110: B. J. Daiuto, T. T. Hartley,
S. P. Chicatelli
The Hyperbolic Map and Applications
to the Linear Quadratic Regulator
VI, 114 pages, 1989

Vol. 111: A. Bensoussan, J. L. Lions (Eds.)
Analysis and Optimization
of Systems
XIV, 1175 pages, 1988.

Vol. 112: Vojislav Kecman
State-Space Models of Lumped
and Distributed Systems
IX, 280 pages, 1988

Vol. 113: M. Iri, K. Yajima (Eds.)
System Modelling and Optimization
Proceedings of the 13th IFIP Conference
Tokyo, Japan, Aug. 31 – Sept. 4, 1987
IX, 787 pages, 1988.

Vol. 114: A. Bermúdez (Editor)
Control of Partial Differential Equations
Proceedings of the IFIP WG 7.2
Working Conference
Santiago de Compostela, Spain, July 6–9, 1987
IX, 318 pages, 1989

Vol. 115: H.J. Zwart
Geometric Theory for Infinite
Dimensional Systems
VIII, 156 pages, 1989.

Vol. 116: M.D. Mesarovic, Y. Takahara
Abstract Systems Theory
VIII, 439 pages, 1989

Lecture Notes in Control and Information Sciences

Edited by M. Thoma and A. Wyner

Vol. 117: K.J. Hunt
Stochastic Optimal Control Theory
with Application in Self-Tuning Control
X, 308 pages, 1989.

Vol. 118: L. Dai
Singular Control Systems
IX, 332 pages, 1989

Vol. 119: T. Başar, P. Bernhard
Differential Games and Applications
VII, 201 pages, 1989

Vol. 120: L. Trave, A. Titli, A. M. Tarras
Large Scale Systems:
Decentralization, Structure Constraints
and Fixed Modes
XIV, 384 pages, 1989

Vol. 121: A. Blaquière (Editor)
Modeling and Control of Systems
in Engineering, Quantum Mechanics,
Economics and Biosciences
Proceedings of the Bellman Continuum
Workshop 1988, June 13–14, Sophia Antipolis, France
XXVI, 519 pages, 1989

Vol. 122: J. Descusse, M. Fliess, A. Isidori,
D. Leborgne (Eds.)
New Trends in Nonlinear Control Theory
Proceedings of an International
Conference on Nonlinear Systems,
Nantes, France, June 13–17, 1988
VIII, 528 pages, 1989

Vol. 123: C. W. de Silva, A. G. J. MacFarlane
Knowledge-Based Control with
Application to Robots
X, 196 pages, 1989

Vol. 124: A. A. Bahnasawi, M. S. Mahmoud
Control of Partially-Known
Dynamical Systems
XI, 228 pages, 1989

Vol. 125: J. Simon (Ed.)
Control of Boundaries and Stabilization
Proceedings of the IFIP WG 7.2 Conference
Clermont Ferrand, France, June 20–23, 1988
IX, 266 pages, 1989

Vol. 126: N. Christopeit, K. Helmes
M. Kohlmann (Eds.)
Stochastic Differential Systems
Proceedings of the 4th Bad Honnef Conference
June 20–24, 1988
IX, 342 pages, 1989

Vol.127: C. Heij
Deterministic Identification
of Dynamical Systems
VI, 292 pages, 1989

Vol. 128: G. Einarsson, T. Ericson,
I. Ingemarsson, R. Johannesson,
K. Zigangirov, C.-E. Sundberg
Topics in Coding Theory
VII, 176 pages, 1989

Vol. 129: W. A. Porter, S. C. Kak (Eds.)
Advances in Communications and
Signal Processing
VI, 376 pages, 1989.

Vol. 130: W. A. Porter, S. C. Kak,
J. L. Aravena (Eds.)
Advances in Computing and Control
VI, 367 pages, 1989

Vol. 131: S. M. Joshi
Control of Large Flexible Space Structures
IX, 196 pages, 1989.

Vol. 132: W.-Y. Ng
Interactive Multi-Objective Programming
as a Framework for Computer-Aided Control
System Design
XV, 182 pages, 1989.

Vol. 133: R. P. Leland
Stochastic Models for Laser Propagation
in Atmospheric Turbulence
VII, 145 pages, 1989.

Vol. 134: X. J. Zhang
Auxiliary Signal Design in Fault
Detection and Diagnosis
XII, 213 pages, 1989

Vol. 135: H. Nijmeijer, J. M. Schumacher (Eds.)
Three Decades of Mathematical System Theory
A Collection of Surveys at the Occasion of the
50th Birthday of Jan C. Willems
VI, 562 pages, 1989

Vol. 136: J. Zabczyk (Ed.)
Stochastic Systems and Optimization
Proceedings of the 6th IFIP WG 7.1
Working Conference,
Warsaw, Poland, September 12–16, 1988
VI, 374 pages. 1989

Lecture Notes in Control and Information Sciences

Edited by M. Thoma and A. Wyner

Vol. 137: S. L. Shah, G. Dumont (Eds.)
Adaptive Control Strategies for Industrial Use
Proceedings of a Workshop
Kananaskis, Canada, 1988
VI, 360 pages. 1989

Vol. 138: D. C. McFarlane, K. Glover
Robust Controller Design Using Normalized Coprime Factor Plant Descriptions
X, 206 pages. 1990

Vol. 139: V. Hayward, O. Khatib (Eds.)
Experimental Robotics I
The First International Symposium
Montreal, June 19-21, 1989
XVII, 613 pages. 1990

Vol. 140: Z. Gajic, D. Petkovski, X. Shen
Singularly Perturbed and Weakly Coupled Linear Control Systems
A Recursive Approach
VII, 202 pages. 1990

Vol. 141: S. Gutman
Root Clustering in Parameter Space
VIII, 153 pages. 1990

Vol. 142: A. N. Gündeş, C. A. Desoer
Algebraic Theory of Linear Feedback Systems with Full and Decentralized Compensators
V, 176 pages. 1990

Vol. 143: H.-J. Sebastian, K. Tammer (Eds.)
System Modelling and Optimization
Proceedings of the 14th IFIP Conference
Leipzig, GDR, July 3-7, 1989
X, 960 pages. 1990

Vol. 144: A. Bensoussan, J. L. Lions (Eds.)
Analysis and Optimization of Systems
Proceedings of the 9th International Conference, Antibes, June 12–15, 1990
XII, 992 pages. 1990

Vol. 145: M. B. Subrahmanyam
Optimal Control with a Worst-Case Performance Criterion and Applications
X, 133 pages. 1990

Vol. 146: D. Mustafa, K. Glover
Minimum Entropy H_∞ Control
IX, 144 pages. 1990

Vol. 147: J. P. Zolésio (Ed.)
Stabilization of Flexible Structures
Third Working Conference,
Montpellier, France, January 1989
V, 327 pages, 1990

Vol. 148: In preparation

Vol. 149: K.H. Hoffmann, W. Krabs (Eds.)
Optimal Control of Partial Differential Equations
Proceedings of the IFIP WG 7.2 International Conference, Irsee, April 9-12, 1990
VI, 245 pages. 1991

Vol. 150: L. C. G. J. M. Habets
Robust Stabilization in the Gap-topology
IX, 126 pages. 1991

Vol. 151: J. M. Skowronski, H. Flashner, R. S. Guttalu (Eds.)
Mechanics and Control
Proceedings of the 3rd Workshop on Control Mechanics in Honor of the 65th Birthday of George Leitmann
January 22-24, 1990, University of Southern California
IV, 497 pages. 1991

Vol. 152: J. D. Aplevich
Implicit Linear Systems
XI, 176 pages. 1991

Vol. 153: O. Hájek
Control Theory in the Plane
X, 272 pages. 1991

Vol. 154: A. Kurzhanski, I. Lasiecka (Eds.)
Modelling and Inverse Problems of Control for Distributed Parameter Systems
Proceedings of IFIP (W.G. 7.2)-IIASA Conference
Laxenburg, Austria, July 24-28, 1989
VII, 179 pages. 1991